INTRODUCTION TO
graph theory

3C½ C

INTRODUCTION TO
graph theory

Robin J. Wilson

OLIVER & BOYD · EDINBURGH

Oliver and Boyd
Tweeddale Court
14 High Street
Edinburgh EH1 1YL
A Division of Longman Group Limited

ISBN 0 05 002534 1

First published 1972

Printed in Great Britain by
Bell & Bain Ltd., Glasgow

Contents

Go forth, my little book! pursue thy way!
Go forth, and please the gentle and the good.
WILLIAM WORDSWORTH

Preface

In the past few years, graph theory has established itself as an important mathematical tool in a wide variety of subjects, ranging from operational research and linguistics to chemistry and genetics; at the same time it has also emerged as a worthwhile mathematical discipline in its own right. For some time there has been a need for an inexpensive introductory text on the subject, suitable both for mathematicians taking courses in graph theory and also for non-specialists wishing to learn the subject as quickly as possible. It is my hope that this book goes some of the way towards filling this need. The only prerequisites to reading it are a basic knowledge of elementary set theory and matrix theory.

The contents of this book may be conveniently divided into four parts. The first of these, consisting of the first four chapters, provides a basic foundation course containing such topics as connectedness, trees, and Eulerian and Hamiltonian paths and circuits. This is then followed by two chapters on planar graphs and colouring, with special reference to problems relating to the four-colour conjecture. The third part (Chapters seven and eight) deals with the theory of directed graphs (digraphs) and with transversal theory, relating these fields to such subjects as Markov chains and network flows. The book ends with a chapter on matroid theory which is intended to tie together the material of the previous chapters as well as to introduce some very recent developments in the subject. The reader who is primarily interested in 'pure' graph theory may choose to omit chapters seven and eight on a first reading; on the other hand, the reader who is mainly concerned with applications may prefer to omit chapters five and six. It is hoped that some of the material in the first three parts may also be found suitable for school sixth forms.

Throughout the book I have attempted to restrict the text to basic material only, using the exercises as a means for introducing further material of lesser importance. The result of this is that there are about 250 exercises, some of which are designed to test understanding of the text, but a large number of which are intended to introduce the reader to new results and ideas. The reader is urged to read through, and become familiar with, every exercise whether or not he works through all of them in detail. The more difficult exercises are indicated by an asterisk (*).

There are several parts of the book which may be omitted on a first reading, either because of their difficulty or because the material they contain is not referred to later in the book; a star ★ is used to designate the beginning and end of such sections. I have used the symbol // to indicate the end (or absence) of a proof, and bold-face type is used for all definitions. Finally, the number of elements in a set S will be denoted throughout by $|S|$, and the empty set will be denoted by $\emptyset$.

It would be impossible to mention by name all of the many people who have helped to make this book possible by their valuable suggestions and comments. I should, however, particularly like to thank John W. Moon, Peter A. Rado and T. D. Parsons for their invaluable help in reading through the final manuscript. I am also grateful to Gian–Carlo Rota and Hallard Croft for introducing me to the subject in the first place. In addition, I should like to thank Mrs Deirdre Concannon for drawing the diagrams and Mrs A. Gillon for typing the manuscript.

Finally I should like to express my thanks: to my students, but for whom this book would have been completed a year earlier; to Mr William Shakespeare and others, for their apt and witty comments at the beginning of each chapter; and most of all to my wife, Joy, for many things which have nothing at all to do with graph theory.

Jesus College, Oxford.   R. J. W.

1 Introduction

*The last thing one discovers in
writing a book is what to put first.*
BLAISE PASCAL

The object of this introductory chapter is to provide (by means of simple examples) an intuitive background to the material to be presented more formally in succeeding chapters. Terms which appear here in bold-face type are to be thought of more as descriptions than as definitions—the idea is that having met the words in an intuitive setting, the reader will not find them totally unfamiliar when he meets them again in more formal surroundings. We advise the reader to read this chapter quickly—and then to forget all about it!

§1. WHAT IS A GRAPH?

Let us begin by considering Figs. 1.1 and 1.2 which depict, respectively, part of an electrical network and part of a road map. It is clear

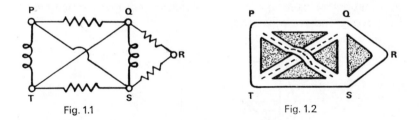

Fig. 1.1   Fig. 1.2

that either of them can be represented diagrammatically by means of points and lines as in Fig. 1.3. The points P, Q, R, S, and T are called **vertices** and the lines are called **edges**; the whole diagram is called a **graph.** (Note that the intersection of the lines PS and QT is not a vertex of the graph since it does not correspond to the meeting of two wires or to a cross-roads.) The **degree** of a vertex is the number of edges which have that vertex as an end-point, and corresponds in Fig. 1.2 to the number of roads at an intersection; thus the degree of the vertex Q is four.

1

Clearly the graph in Fig. 1.3 can also represent other situations. For example, if *P*, *Q*, *R*, *S* and *T* represent football teams, then the existence of an edge might correspond to the playing of a game between the teams at its endpoints (so that in Fig. 1.3, *P* has played against *S* but not against *R*); in this case, the degree of a vertex is the number of games played by the corresponding team.

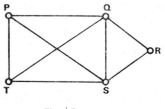

Fig. 1.3

An alternative way of depicting the above situations is given by the graph in Fig. 1.4. Here we have removed the 'crossing' of the lines *PS* and *QT* by drawing the line *PS* outside the rectangle *PQST*. Note that the resulting graph still tells us how the electrical network is wired up, whether there is a direct road from one intersection to another, and which football teams have played which; the only information we have lost concerns 'metrical' properties (length of road, straightness of wire, etc.).

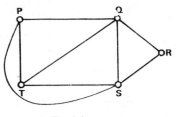

Fig. 1.4

The point we are trying to make is that a graph is a representation of a set of points and of the way they are joined up, and that for our purposes any metrical properties are irrelevant. From this point of view, any two graphs which represent the same situation (such as the ones shown in Figs. 1.3 and 1.4) will be regarded as essentially the same graph. More precisely, we shall say that two graphs are **isomorphic** if there is a one-one correspondence between their vertices which has the property that two vertices are joined by an edge in one graph if and only if the corresponding vertices are joined by an edge in the other. Another graph isomorphic to the graphs in Figs. 1.3 and 1.4 is shown in Fig. 1.5; note that in this graph all idea

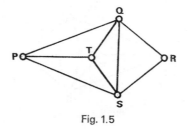

Fig. 1.5

of space and distance has gone, although we can still tell at a glance which points are joined by a wire or a road.

It is worth pointing out that the graph we have been discussing so far is a particularly 'simple' graph, in the sense that there is never more than one edge joining a given pair of vertices. Suppose, now, that in Fig. 1.5 the roads joining Q and S, and S and T have too much traffic to carry; then the situation could be eased by building extra roads joining these points, and the resulting diagram would look like Fig. 1.6. (The edges joining Q and S, or S and T, are called **multiple edges**.) If in addition we wish to build a car park at

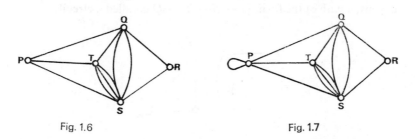

Fig. 1.6   Fig. 1.7

P, then this could be indicated on the graph by drawing an edge from P to itself, usually called a **loop** (see Fig. 1.7). In this book, a graph will in general contain loops and multiple edges; graphs containing no loops or multiple edges (such as the graph in Fig. 1.5) will be referred to as **simple graphs.**

The study of **directed graphs** (or **digraphs,** as we shall usually abbreviate them) arises out of the question, 'what happens if all of the roads are one-way streets?' An example of a digraph is given in Fig. 1.8, the directions of the one-way streets being indicated by

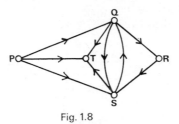

Fig. 1.8

arrows; (in this particular example, there would be utter chaos at *T*, but that does not stop us from studying such situations!) Note that if *not all* of the streets are one-way, then we can obtain a digraph by drawing for each two-way road two directed edges, one in each direction. We shall be discussing digraphs in some detail in Chapter VII.

Much of graph theory involves the study of paths of various kinds, a **path** being essentially a sequence of edges, one following on after another; thus, for example, in Fig. 1.5 $P \to Q \to R$ is a 'way of getting from *P* to *R*' and is a path of length two, and similarly $P \to S \to Q \to T \to S \to R$ is a path of length five. For obvious reasons, a path of the form $Q \to S \to T \to Q$ is called a **circuit.**

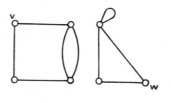

Fig. 1.9

In general, given two vertices *v* and *w* in a graph, it is not always possible to find a path connecting them (see Fig. 1.9); such a path will exist only when the graph is 'in one piece'. We can make this clearer by considering the graph whose vertices are the stations of the London Underground and the New York Subway, and whose edges are the various lines joining them; it is obviously impossible to get from Charing Cross to Grand Central Station using only edges of the graph. On the other hand, if we confine our attention to the stations and lines of the London Underground, then we can get from any station to any other. A graph in which any two vertices are connected by a path is called a **connected graph;** such graphs will be discussed in Chapter III.

Much of Chapters III and IV will be devoted to the study of graphs containing a path or paths having some particular property. In Chapter III, for example, we shall be discussing graphs which

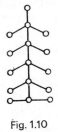

Fig. 1.10

contain paths which include every edge or every vertex exactly once, ending up at the initial vertex; such graphs will be called **Eulerian** and **Hamiltonian** graphs respectively. For example, the graph in Fig. 1.5 is Hamiltonian (a possible path being $P \to Q \to R \to S \to T \to P$) but is not quite Eulerian, since any path which includes every edge exactly once (e.g. $P \to Q \to T \to P \to S \to R \to Q \to S \to T$) must end up at a vertex different from the initial vertex.

We shall also be interested in connected graphs in which there is only one path connecting each pair of vertices; such graphs are called **trees** (generalizing the idea of a family tree) and will be considered in Chapter IV. We shall see that a tree can be defined as a connected graph which contains no circuits (see Fig. 1.10).

To change the subject a little, the reader will recall that when we were discussing Fig. 1.3, we pointed out that there were graphs (e.g. Figs. 1.4 and 1.5) which were isomorphic to the graph under consideration but which contained no crossings. Any graph which can be redrawn in this way without crossings is called a **planar graph**, such graphs being of the utmost importance in the theory of graphs. In Chapter V we shall give several criteria for planarity, some of which will involve the properties of subgraphs of the graph in question, and others of which will involve the fundamental notion of **duality.**

Planar graphs also play an important rôle in colouring problems; to motivate such problems, let us return to our 'road-map' graph, and let us suppose that Shell, Esso, B.P., and Gulf wish to put up five garages between them at P, Q, R, S and T. Let us further assume that for economic reasons no company wishes to erect two garages at neighbouring corners. Then one solution would be for Shell to build at P, Esso to build at Q, B.P. at S, and Gulf at T, leaving either Shell or Gulf to build at R; however, if Gulf decides to back out of the whole agreement, then it is clearly impossible for the other three companies to erect the garages in the specified manner.

This problem will be discussed in more colourful language in Chapter VI, where we investigate the question of whether the vertices of a given simple graph can be coloured using k given colours in such a way that every edge of the graph has endpoints of different colours. If the graph happens to be planar, then we shall see that it is always possible to colour its vertices in the above-mentioned way if five colours are available; moreover, it is conjectured (although this has never been proved) that the same is true if only four colours are available—this is the famous **four-colour conjecture.** (A possibly more familiar version of this conjecture is that if we have a map with several countries on it, then it is always possible to colour the countries of the map with four colours in such a way that no two neighbouring countries share the same colour.)

In Chapter VIII we shall investigate various combinatorial problems, including the celebrated **'marriage problem'** which asks under what conditions a collection of boys, each of whom knows several girls, can be married off in such a way that each boy marries a girl he knows. This problem can be easily expressed in the language of transversal theory, a very important branch of combinatorial mathematics which we discuss in **§26.** It will turn out that these topics are closely related to the problem of finding the number of paths connecting two given vertices in a graph or digraph, subject to the restriction that no two of the paths have an edge in common.

We conclude Chapter VIII with a discussion of network flows and transportation problems. To describe these problems, we suppose that Fig. 1.5 represents part of an electrical network made up of wires of different materials; the problem is then to find out how large a current can safely be passed through the entire circuit from P to R, given the various currents which each separate wire can take without burning out. Alternatively, we can think of P as a factory and R as a market and the edges of the graph as various channels through which goods can be sent; in this case we want to know how much can be sent from the factory to the market, given the capacities of the various channels.

We end the book with a chapter on the theory of matroids; this chapter is intended to tie together the material of the previous chapters as well as to satisfy the maxim 'be wise—generalize!' In fact **matroid theory** is simply the study of sets with 'independence structures' defined on them, generalizing not only properties of linear independence in vector spaces but also several of the results in graph theory obtained earlier in the book. However, as we shall see, matroid theory is far from being 'generalization for generalization's sake'; on the contrary, it gives us a deeper insight into several graph-theoretical problems as well as including among its applications simple proofs of results in transversal theory which are awkward to prove by more traditional methods. We believe that matroid theory has an important rôle to play in the development of combinatorial theory in the coming years, and we have included it in our book for this reason.

We hope that this introductory chapter has been useful to our readers in setting the stage and describing some of the things which lie ahead; we now embark upon a formal treatment of the subject.

Exercises

(*1a*) Show how the following may be regarded as graphs or digraphs: (*i*) the vertices and edges of a polyhedron; (*ii*) the plan of a maze; (*iii*) the friendships between people at a party; (*iv*) the Atomium at Brussels; (*v*) the stages of play in noughts-and-crosses (tic-tac-toe); (*vi*) a family tree; (*vii*) a tennis tournament; (*viii*) the divisors of the number 60; (*ix*) the countries on a map; (*x*) the layout of the exhibits in an exhibition.

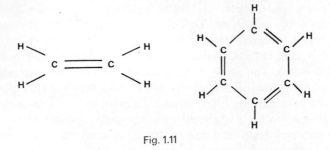

Fig. 1.11

(*1b*) Fig. 1.11 represents the molecules of ethylene and benzene, C and H denoting carbon and hydrogen atoms respectively. Would you regard these diagrams as graphs? If so, can you suggest some necessary conditions that we should impose on a graph in order that it should represent a hydrocarbon?

(*1c*) Think up five more applications of graph theory (as varied as possible).

2 *Definitions and Examples*

I hate definitions!
BENJAMIN DISRAELI

In this chapter, the foundations are laid for a proper study of graph theory. §2 formalizes some of the basic definitions mentioned in Chapter I, and §3 provides a variety of examples. Diagrams are used throughout to clarify the material, and the justification for their use is given in §4. A description of some typical applications of the theory is deferred until we have more machinery at our disposal (§11).

§2. DEFINITIONS

We shall begin by defining a **simple graph** G to be a pair $(V(G), E(G))$, where $V(G)$ is a non-empty finite set of elements called **vertices** (or **nodes,** or **points**), and $E(G)$ is a finite set of unordered pairs of distinct elements of $V(G)$ called **edges** (or **lines**); $V(G)$ is sometimes called the **vertex-set** and $E(G)$ the **edge-set** of G. For example, Fig. 2.1 represents the simple graph G whose vertex-set $V(G)$ is the

set
ordered/
unordered
pair
finiteness
integers

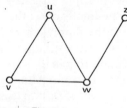

Fig. 2.1

set $\{u, v, w, z\}$, and whose edge-set $E(G)$ consists of the pairs $\{u, v\}$, $\{v, w\}$, $\{u, w\}$ and $\{w, z\}$. The edge $\{v, w\}$ is said to **join** the vertices v and w; note that since $E(G)$ is a set, rather than a family[†], there can never be more than one edge joining a given pair of vertices of a simple graph.

[†] We use the word 'family' to mean a collection of elements, some of which may occur several times; for example, $\{a, b, c\}$ is a set, but (a, a, a, b, c, c) is a family.

B

It turns out that many of the results which can be proved about simple graphs may be extended without difficulty to more general objects in which two vertices may have more than one edge joining them. In addition, it is often convenient to remove the restriction that any edge must join two *distinct* vertices, and to allow the existence of **loops,** i.e. edges joining vertices to themselves. The resulting object, in which loops and multiple edges are allowed, is then called a **general graph**—or, simply, a **graph** (see Fig. 2.2). We emphasize the fact that every simple graph is a graph, but not every graph is a simple graph.

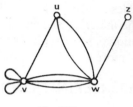

Fig. 2.2

More formally, a **graph** is defined to be a pair $(V(G), E(G))$, where $V(G)$ is a non-empty finite set of elements called **vertices,** and $E(G)$ is a finite family of unordered pairs of (not necessarily distinct) elements of $V(G)$ called **edges;** note that the use of the word 'family' permits the existence of multiple edges. We shall call $V(G)$ the **vertex-set** and $E(G)$ the **edge-family** of G; thus in Fig. 2.2, $V(G)$ is the set $\{u, v, w, z\}$ and $E(G)$ is the family consisting of the edges $\{u, v\}$, $\{v, v\}$, $\{v, v\}$, $\{v, w\}$, $\{v, w\}$, $\{v, w\}$, $\{u, w\}$, $\{u, w\}$ and $\{w, z\}$. Any edge of the form $\{v, w\}$ is said to **join** the vertices v and w; thus every loop $\{v, v\}$ joins the vertex v to itself. Although in this book we shall sometimes have to restrict ourselves to simple graphs, we shall wherever possible prove our results for graphs in general.

A subject related to graph theory is the study of digraphs (sometimes called directed graphs or networks, although we shall be using the word 'network' in a slightly different sense). A **digraph** D is defined to be a pair $(V(D), A(D))$, where $V(D)$ is a non-empty finite set of elements called **vertices,** and $A(D)$ is a finite family of *ordered* pairs of elements of $V(D)$ called **arcs** (or **di-edges**). An arc whose first element is v and whose second element is w is called an **arc from v to w** and is written (v, w); note that two arcs of the form (v, w) and

(w, v) are different. Fig. 2.3 represents a digraph whose arcs are (u, v), (v, v), (v, w), (v, w), (w, v), (w, u) and (w, z), the ordering of the vertices in an arc being indicated by an arrow.

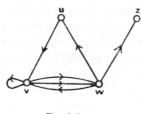

Fig. 2.3

Digraphs will be studied in further detail in Chapter VII; in the meantime, we shall be content to point out that although graphs and digraphs are essentially different objects, a graph can in certain circumstances be thought of as a digraph in which there are two arcs, one in each direction, corresponding to each edge (see Fig. 2.4).

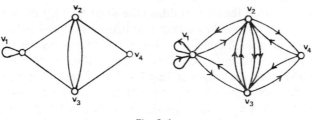

Fig. 2.4

REMARK ON TERMINOLOGY. The language of graph theory is decidedly non-standard—every author has his own terminology. In this book we are using essentially the terminology of Busacker and Saaty.[2] Several graph theorists, however, use the term 'graph' to mean what we have called a simple graph. It is also common, especially when discussing applications, to see the word 'graph' used for what we have called a digraph. To make matters worse, one sometimes sees the term 'graph' used for the object which results if, in the definition of a graph, we remove the restriction that the vertex-set and edge-family must both be finite. (If they are in fact both infinite, then we get what we call an **infinite graph**; we defer a study of infinite graphs until §8, although they will appear in exercises before then.) It should be emphasized that any of the above definitions of a graph is perfectly valid, provided that one is always consistent; we repeat that in this book, *all graphs are finite and undirected, loops and multiple edges being allowed unless specifically excluded.*

Before giving examples of some important types of graph (in §3), it will be convenient to introduce a few more simple definitions.

Two vertices v and w of a graph G are said to be **adjacent** if there is an edge joining them (i.e. there is an edge of the form $\{v, w\}$); the vertices v and w are then said to be **incident** to such an edge. Similarly, two distinct edges of G are **adjacent** if they have at least one vertex in common. The **degree** (or **valency**) of a vertex v of G is the number of edges incident to v, and is written $\rho(v)$; in calculating the degree of a vertex v, we shall (unless otherwise stated) make the convention that a loop at v contributes two (rather than one) to the degree of v. Any vertex of degree zero is called an **isolated vertex,** and a vertex of degree one is a **terminal vertex** (or **endpoint**). Thus the graph in Fig. 2.2 has one terminal vertex, one vertex of degree three, one of degree six and one of degree eight.

It is easy to see that if we add up the degrees of all the vertices of a graph, the result will be an even number—in fact, twice the number of edges—since each edge contributes exactly two to the sum. This result, known two hundred years ago to Euler, is often called the **'handshaking lemma'** since it implies that if several people shake hands, the total number of hands shaken must be even—precisely because two hands are involved in each handshake. An immediate corollary of the handshaking lemma is that in any graph the number of vertices of odd degree must be even. The analogue of this result for digraphs will be presented in §23.

functions
injective
surjective   Two graphs G_1 and G_2 are **isomorphic** if there is a one-one correspondence between the vertices of G_1 and those of G_2 with the property that the number of edges joining any two vertices of G_1 is equal to the number of edges joining the corresponding vertices of G_2. Thus the two graphs shown in Fig. 2.5 are isomorphic under the correspondence $u \leftrightarrow l, v \leftrightarrow m, w \leftrightarrow n, x \leftrightarrow p, y \leftrightarrow q, z \leftrightarrow r$; note that there are only six vertices—the other points at which edges cross are not vertices. A **subgraph** of a graph G is simply a graph, all of whose vertices belong to $V(G)$ and all of whose edges belong to $E(G)$; thus the graph in Fig. 2.1 is a subgraph of the graph in Fig. 2.4, but is not

a subgraph of either graph in Fig. 2.5 (since the latter graphs contain no 'triangle').

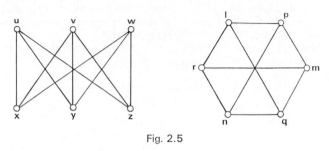

Fig. 2.5

Finally, if G is a graph with vertex-set $\{v_1, \ldots, v_n\}$, the **adjacency** ~matrix~ **matrix** of G (corresponding to the given labelling of the vertices) is the $n \times n$ matrix $A = (a_{ij})$, in which a_{ij} is the number of edges of G joining v_i and v_j; for example, the matrix in Fig. 2.6 is the adjacency

$$\begin{pmatrix} 1 & 1 & 1 & 0 \\ 1 & 0 & 2 & 1 \\ 1 & 2 & 0 & 1 \\ 0 & 1 & 1 & 0 \end{pmatrix}$$

Fig. 2.6

matrix of the graph shown in Fig. 2.4. We can clearly get several different adjacency matrices from a given graph by relabelling the vertices—this will correspond to changing the order of the rows and columns of A—but the result will always be a symmetric matrix of non-negative integers with the property that the sum of the entries in any row or column is the degree of the corresponding vertex (where, this time, each loop contributes only one to the degree). Conversely, given any symmetric matrix, all of whose entries are non-negative integers, we can easily construct the graph (unique up to isomorphism) which has the given matrix as its adjacency matrix. It follows that graph theory may be regarded as essentially the study of a particular type of matrix. (We shall see in exercise *2g* that graph theory may also be regarded as the study of a particular type of polynomial.)

Exercises

(2a)   Let G_n be the graph with vertex-set $\{v_1, ..., v_n\}$ in which the vertices v_i and v_j are adjacent if and only if i and j are relatively prime; draw the graphs G_4 and G_8 and find their adjacency matrices. Show also that if $m < n$, then G_m is a subgraph of G_n.

(2b)   Show that the two graphs in Fig. 2.7 are isomorphic, but that the two graphs in Fig. 2.8 are not isomorphic.

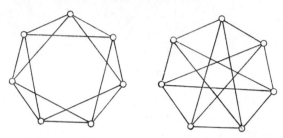

Fig. 2.7

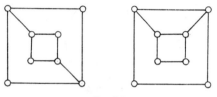

Fig. 2.8

(2c)   Show that up to isomorphism there are exactly four simple graphs on three vertices, and eleven on four vertices; how many are there on five vertices? (See the appendix.)

(2d)   How would you define the concept of isomorphism between digraphs? Check that your definition agrees with the one given in §22, and show that up to isomorphism there are exactly sixteen simple digraphs on three vertices.

(2e)?   Let G be a simple graph with at least two vertices; show that G contains two vertices of the same degree.

(2f)   The **incidence matrix** of a simple graph with vertex-set $\{v_1, ..., v_n\}$ and edge-set $\{e_1, ..., e_m\}$ is the $m \times n$ matrix $A = (a_{ij})$, in which $a_{ij} = 1$ if vertex v_j and edge e_i are incident, and $a_{ij} = 0$ otherwise. Show that the sum of the entries in any column is the degree of the corresponding vertex; what can you say about the sum of the entries in any row? Also show how one can define an incidence matrix for a (general) graph.

(2g)   Let G be a graph with vertex-set $\{v_1, ..., v_n\}$, and let $f(G)$ be the polynomial in variables $x_1, ..., x_n$ defined by

$$f(G) = x_1^{\sigma_1} x_2^{\sigma_2} ... x_n^{\sigma_n} \Pi(x_j - x_i)^{\alpha_{ij}},$$

where the product extends over all pairs of integers $i < j$, α_{ij} denotes the number of edges joining v_i and v_j, and σ_i denotes the number of loops at v_i. Show that G is determined up to isomorphism by the polynomial $f(G)$. To which graphs do factors of $f(G)$ correspond? Illustrate your answer using the graph shown in Fig. 2.4.

(2h) ? The **line graph** $L(G)$ of a simple graph G is the graph whose vertices are in one-one correspondence with the edges of G, two vertices of $L(G)$ being adjacent if and only if the corresponding edges of G are adjacent. Show that the line graphs of the two graphs in Fig. 2.9

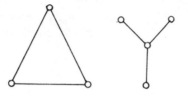

Fig. 2.9

are isomorphic; also find an expression for the number of edges of $L(G)$ in terms of the degrees of the vertices of G.

(*2i)   A **k-graph** is defined to be a pair $(V(G), E(G))$, where $V(G)$ is as usual, and $E(G)$ is a finite family of unordered k-tuples of elements of $V(G)$. Verify that a graph is a 2-graph, and show how the definitions of such terms as 'incidence', 'adjacency', 'degree' and 'terminal vertex' may be extended to k-graphs.

(*2j)   If G is a simple graph with edge-set $E(G)$, the **vector space associated with G** is defined to be the vector space over the field of integers modulo two whose elements are subsets of $E(G)$, the sum $E \oplus F$ of two sets E, F of edges being defined as the set of all edges which lie either in E or in F but not both, and scalar multiplication being defined in the obvious way (i.e. $1.E = E$, $0.E = \varnothing$). Show that this does in fact define a vector space, and describe a basis for it. Illustrate these ideas with reference to the graph of Fig. 2.1.

§3. EXAMPLES OF GRAPHS

In this section we shall examine some important types of graph; the reader is advised to become familiar with them since they will appear frequently in examples and exercises.

NULL GRAPHS. A graph whose edge-set is empty is called a **null graph** (or **totally-disconnected graph**). We shall denote the null graph on n vertices by N_n; N_4 is shown in Fig. 3.1. Note that in a null graph, every vertex is isolated. Null graphs are not very interesting.

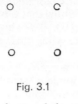

Fig. 3.1

COMPLETE GRAPHS. A simple graph in which every pair of distinct vertices are adjacent is called a **complete graph.** The complete graph on n vertices is usually denoted by K_n, K_4 and K_5 being shown in Figs. 3.2 and 3.3. The reader should check that K_n has exactly $\frac{1}{2}n(n-1)$ edges.

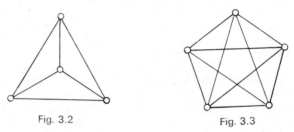

Fig. 3.2   Fig. 3.3

REGULAR GRAPHS. A graph in which every vertex has the same degree is called a **regular graph;** if every vertex has degree r, the graph is called **regular of degree r.** Of particular importance in colouring problems (to be discussed in Chapter VI) are the **cubic** (or **trivalent**) graphs which are regular graphs of degree three (for example, Figs. 2.5, 3.2 and 3.4); another well-known example of a

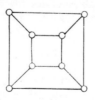
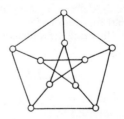

Fig. 3.4   Fig. 3.5

cubic graph is the so-called **Petersen graph** shown in Fig. 3.5. Note that every null graph is regular of degree zero, and that the complete graph K_n is regular of degree $n - 1$.

PLATONIC GRAPHS. Of special interest among the regular graphs are the so-called **Platonic graphs,** the graphs formed by the vertices and edges of the five regular (Platonic) solids—the tetrahedron, ~~solid~~ cube, octahedron, dodecahedron and icosahedron. The tetrahedral ~~geom.~~ graph is K_4 (see Fig. 3.2), and the graphs of the cube and octahedron are shown in Fig. 3.4 and 3.6; the dodecahedral graph will appear in Fig. 7.4 (see page 37). We leave the drawing of the icosahedral graph as an exercise for the reader.

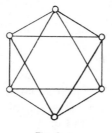

Fig. 3.6

BIPARTITE GRAPHS. Suppose that the vertex-set of a graph G can be divided into two disjoint sets V_1 and V_2, in such a way that every edge of G joins a vertex of V_1 to a vertex of V_2 (see Fig. 3.7); G is

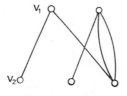

Fig. 3.7

then said to be a **bipartite graph** (sometimes denoted by $G(V_1, V_2)$ if we wish to specify the two sets involved). An alternative way of thinking of a bipartite graph is in terms of colouring its vertices with two colours, say red and blue—a graph is bipartite if we can colour each vertex red or blue in such a way that every edge has a red end and a blue end. It is worth emphasizing that in a bipartite graph $G(V_1, V_2)$, it is not necessarily true that every vertex of V_1 is joined to every vertex of V_2; if however this does happen, and if G is simple, then G is called a **complete bipartite graph**, usually denoted by $K_{m,n}$ where m and n are the numbers of vertices in V_1 and V_2 respectively. For example, Fig. 3.8 represents $K_{4,3}$, and two drawings

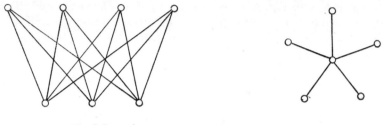

Fig. 3.8   Fig. 3.9

of $K_{3,3}$ appeared in Fig. 2.5. Note that $K_{m,n}$ has $m+n$ vertices and mn edges. A complete bipartite graph of the form $K_{1,n}$ is called a **star graph**, $K_{1,5}$ being shown in Fig. 3.9.

THE UNION AND SUM OF TWO GRAPHS. There are several ways of combining two graphs to make a larger graph; we shall illustrate two of these. If the two graphs are taken to be $G_1 = (V(G_1), E(G_1))$ and $G_2 = (V(G_2), E(G_2))$, where $V(G_1)$ and $V(G_2)$ are assumed to be

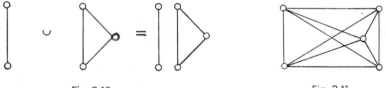

Fig. 3.10   Fig. 3.11

disjoint, then their **union** $G_1 \cup G_2$ is defined as the graph with vertex-set $V(G_1) \cup V(G_2)$ and edge-family $E(G_1) \cup E(G_2)$ (see Fig. 3.10). We can also form the **sum** of G_1 and G_2 (denoted by $G_1 + G_2$) by taking the union of G_1 and G_2 and drawing an edge from each vertex of G_1 to each vertex of G_2; for example, the graph $K_2 + K_3$ is shown in Fig. 3.11. Note that $K_{m,n}$ could have been defined as the sum of N_m and N_n. The reader should check that the union and sum operations can be extended to any finite number of graphs, and that they are commutative and associative.

CONNECTED GRAPHS. As the reader has probably noticed, almost all the graphs we have discussed so far have been 'in one piece', the main exceptions being the null graphs $N_n (n \geq 2)$ and the union of graphs, both of which consist of 'bits which are not joined together'. We can formalize this distinction by defining a graph to be **connected** if it cannot be expressed as the union of two graphs; otherwise it is **disconnected.** It is clear that any disconnected graph G can be expressed as the union of a finite number of connected graphs—each of these connected graphs is called a (**connected**) **component** of G. (A graph with three components is shown in Fig. 3.12.) When proving

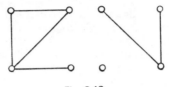

Fig. 3.12

results about graphs in general, it is often possible and convenient to obtain the corresponding results for connected graphs, and then apply them to each component separately.

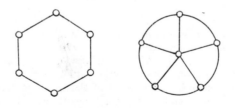

Fig. 3.13

CIRCUIT GRAPHS AND WHEELS. A connected graph which is regular of degree two is called a **circuit graph,** the circuit graph on n vertices being denoted by C_n. The sum of N_1 and C_{n-1} ($n \geq 3$) is called the **wheel** on n vertices, and is written W_n. Fig. 3.13 shows C_6 and W_6; W_4 appeared in Fig. 3.2.

THE COMPLEMENT OF A SIMPLE GRAPH. Let G be a simple graph with vertex-set $V(G)$. The **complement** $\bar{G}$ of G is the simple graph which has $V(G)$ as its vertex-set, and in which two vertices are adjacent if and only if they are not adjacent in G. It follows that if G has n vertices, then $\bar{G}$ can be constructed by removing from K_n all the edges of G, (G being regarded as a subgraph of K_n). Note that the complement of a complete graph is a null graph, and vice versa; the complement of a regular graph is regular.

Exercises

(*3a*) Draw all cubic graphs with at most eight vertices.

(*3b*) Give examples (where they exist) of (*i*) a bipartite graph which is regular; (*ii*) a cubic graph on nine vertices; (*iii*) a Platonic graph which is bipartite; (*iv*) (for each n) a simple graph with n vertices and $\frac{1}{2}(n-1)(n-2)$ edges; (*v*) a simple graph which is isomorphic to its line graph; (*vi*) a simple graph whose complement and line graph are isomorphic; (*vii*) a wheel whose complement is a circuit graph; (*viii*) four connected graphs which are regular of degree four; (*ix*) a Platonic graph which is the line graph of another Platonic graph; (*x*) a connected graph which cannot be expressed as the sum of two graphs.

(*3c*) What can you say about (*i*) the sum of two complete graphs; (*ii*) the complement of a complete bipartite graph; (*iii*) the complements of the tetrahedral, cube and octahedral graphs; (*iv*) the complement of the sum of two simple graphs?

(*3d*) Let G, H and K be simple graphs; prove or disprove the following:

(*i*) $G \cup (H+K) = (G \cup H)+(G \cup K)$;

(*ii*) $G+(H \cup K) = (G+H) \cup (G+K)$.

(*3e*) Describe the adjacency matrices of complete graphs, null graphs, bipartite graphs and circuit graphs; what can you say about the adjacency matrices of a simple graph and its complement?

(*3f*) Show how a bipartite graph can be used to depict (*i*) the friendships between a set of boys and a set of girls; (*ii*) the markets to which various factories send their goods. Suggest three further applications of bipartite graphs.

(*3g*) Let G be a bipartite graph whose largest vertex-degree is p. Show that there exists a bipartite graph $G' = G'(V_1, V_2)$ which is regular of degree p, which contains G as a subgraph, and in which V_1 and V_2 contain the same number of vertices.

(*3h*) Show that the line graph of K_n has $\frac{1}{2}n(n-1)$ vertices and is regular of degree $2n-4$; obtain similar results for $K_{m,n}$. Show also that a simple graph is isomorphic to its line graph if and only if it is regular of degree two, and describe all such graphs.

(3i)   A simple graph which is isomorphic to its complement is called **self-complementary**. Show that the number of vertices of a self-complementary graph must be of the form $4k$ or $4k+1$ where k is an integer, and find all self-complementary graphs on four and five vertices.

(3j)   Show that in a gathering of six people, there are either three people who all know each other of there are three people none of whom knows either of the other two.

(3k)   Give examples of (i) an infinite graph with infinitely many terminal vertices; (ii) an infinite graph with uncountably many vertices and edges; (iii) an infinite connected cubic graph; (iv) an infinite bipartite graph; (v) an infinite graph which is isomorphic to its line graph.

(*3l)   An **automorphism** ϕ of a simple graph G is a one-one mapping of the vertex-set of G onto itself with the property that $\phi(v)$ and $\phi(w)$ are adjacent if and only if v and w are. Show that (i) the automorphisms of G form a group under composition (the **automorphism group** $\Gamma(G)$ of G); (ii) the groups $\Gamma(G)$ and $\Gamma(\overline{G})$ are isomorphic; (iii) $\Gamma(K_n)$ is the symmetric group on n elements; (iv) $\Gamma(C_n)$ is the dihedral group of order $2n$; (v) if P denotes the Petersen graph, then $\Gamma(P)$ is the symmetric group on five elements. Find the automorphism group of $K_{m,n}$, and give an example of a graph whose automorphism group is cyclic of order three.

(*3m)   The **eigenvalues of a graph** are defined to be the eigenvalues of its adjacency matrix; find the eigenvalues of C_5 and W_5. Show that the sum of the eigenvalues of a simple graph is zero, and illustrate this in the case of complete graphs and complete bipartite graphs.

§4. EMBEDDINGS OF GRAPHS

Up to now we have been using diagrams to represent graphs, a vertex being represented by a point or small circle, and an edge by a line or curve. Such diagrams are very useful for investigating the properties of particular graphs, and it is natural to ask what it actually means to 'represent' a graph by a diagram, and whether all graphs can be so represented. The reader who is quite happy drawing pictures and is not concerned with the justification for doing so may omit this section for the time being, secure in the knowledge that everything he does is all right—but he will need to refer back here when he reaches Chapter V.

★What we should like to be able to do is to draw graphs in some space—the plane or Euclidean 3-space, for example—in such a way that there are no 'crossings', (a term which will be defined formally later on, but whose intuitive meaning is clear). For example, Fig. 4.1

Fig. 4.1

represents K_4, but it contains a crossing; we may wish to find a representation (e.g. Fig. 3.2) which contains no crossings. We shall see in fact that every graph can be drawn without crossings in 3-space, but that such a drawing is *not* always possible in the plane; in particular, as we shall show in §**12**, K_5 and $K_{3,3}$ (Figs. 3.3 and 2.5) cannot be drawn in the plane without crossings.

Before defining an embedding of a graph, we remind the reader that a **Jordan curve** (or **Jordan arc**) in the plane is a continuous curve which does not intersect itself; a **closed Jordan curve** is one whose endpoints coincide (see Fig. 4.2). Jordan curves can similarly be

Fig. 4.2

defined in 3-space, or on the surface of such bodies as the sphere and the torus. Later on, we shall be using a form of the famous **Jordan curve theorem** which states that if $\mathscr{C}$ is a closed Jordan curve in the plane, and x and y are any two distinct points of $\mathscr{C}$, then any Jordan curve connecting x and y must either lie completely inside $\mathscr{C}$ (except, of course, for the points x and y), lie completely outside $\mathscr{C}$ (with the same exceptions), or intersect $\mathscr{C}$ at some point other than x and y (see Fig. 4.3). (For further details about the Jordan curve theorem

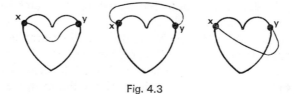

Fig. 4.3

and related topics, see any decent analysis book, for example that by Apostol.[13])

We are now ready to define an embedding of a graph in a given space; the spaces we have in mind are those in which Jordan curves can be defined, but we shall be primarily concerned with the plane and 3-space. A graph G **can be embedded** (or **has an embedding**) in a given space if it is isomorphic to a graph drawn in the space with points representing vertices of G and Jordan curves representing edges in such a way that thre are no crossings; in this definition, a **crossing** is said to occur if either (i) the Jordan curves corresponding to two edges intersect at a point which corresponds to no vertex, or (ii) the Jordan curve corresponding to an edge passes through a point which corresponds to a vertex which is not incident to that edge. (Case (ii) is illustrated in Fig. 4.4; note that the vertex v is not incident to the edge e_1.)

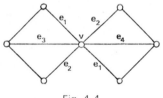

Fig. 4.4

We shall now prove the principal result of this section—that every graph can be drawn without crossings in 3-space. The corresponding result for infinite graphs containing 'not too many' vertices and edges is given in exercise *4a*; note that the theorem is untrue for a general infinite graph.

THEOREM 4A. *Every graph can be embedded in Euclidean 3-space.*
Proof. We shall give an explicit construction for the embedding. First place the vertices of the graph at distinct points of the x-axis; then for each edge, choose a plane through the x-axis in such a way that distinct edges of the graph correspond to distinct planes. (This can always be done since there are only finitely many edges.)

The desired embedding is then obtained as follows: for each loop of the graph we draw in the corresponding plane a circle passing through the relevant vertex; for each edge joining two distinct

Cartesian
co-ords
in
$\mathbb{R}^3$.

vertices we draw in the corresponding plane a semicircle connecting these two vertices. Clearly none of these curves can intersect since they lie in different planes; the result now follows immediately.//

The above theorem gives us the justification we were seeking for using diagrams to depict graphs; we simply take a three-dimensional representation and project it down onto the plane, making sure that no two vertices are projected into the same point. In general, of course, such a method will lead to crossings, but in some cases we will get diagrams with no crossings. This can arise only when the graph in question can be embedded in the plane; such a graph is called a **planar graph.** Planar graphs will be studied in some detail in Chapter V, but we have met several examples already, e.g. K_4, the null graphs, the Platonic graphs, the circuit graphs, the wheels and the star graphs.

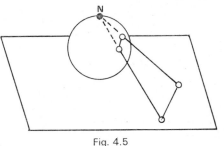

Fig. 4.5

We conclude this section by proving a simple result which will be needed later on. The proof will involve the following definition: if G is a graph embedded in some space, then a point x of the space is said to be **disjoint from** G if x represents neither a vertex of G nor a point which lies on an edge of G.

THEOREM 4B. *A graph is planar if and only if it can be embedded on the surface of a sphere.*

Proof. Let G be a graph embedded on the surface of a sphere; place the sphere on a plane in such a way that the 'north pole' (the point diametrically opposite the point of contact) is disjoint from G. The desired planar representation is then obtained by stereographic projection from the north pole (see Fig. 4.5). The converse is similar and will be left as an exercise.//

Exercises

(*4a*)   Show that an infinite graph can be embedded in Euclidean 3-space if and only if its vertex-set and edge-set can each be put in one-one correspondence with a subset of the real numbers.

(*4b*)   Verify that the above-mentioned examples of planar graphs are in fact planar. Show that a subgraph of a planar graph is planar, and, assuming that $K_{3,3}$ is non-planar, determine which complete bipartite graphs are planar.

(*4c*)   Show that any simple graph can be embedded in Euclidean 3-space in such a way that all of its edges are represented by straight lines.

(*4d*)   Verify theorem 4B in the case of the Platonic graphs.

(*4e*)   Verify that K_5 and $K_{3,3}$ can each be embedded on the surface of a torus. (The torus is illustrated in Fig. 14.1 on page 70).

(**4f*)   Show how a k-graph (see exercise *2i*) can be embedded in Euclidean $(k+1)$-space.

(**4g*)   The **dimension** of a graph G is the smallest integer k with the property that G can be embedded in Euclidean k-space in such a way that if two vertices v and w are adjacent, then the distance between them is one. Show that the dimension of a graph always exists, and find the dimensions of (*i*) K_n; (*ii*) the circuit graphs; (*iii*) the Platonic graphs. ★

3 Paths and Circuits

... So many paths that wind and wind,
While just the art of being kind
Is all the sad world needs.

ELLA WHEELER WILCOX

Now that we have a reasonable armoury of graphs at our disposal, we can start looking at their properties. In order to do this, we need some definitions which describe ways of 'going from one vertex to another'. In §5, we shall give these definitions, and prove some results on connectedness. §6 and §7 are devoted to a rather more detailed study of two particular types of graph, those which contain paths which include every edge, and those which contain circuits which include every vertex. We conclude the chapter with a section on infinite graphs.

§5. MORE DEFINITIONS

Given any graph G, an **edge-sequence** in G is a finite sequence of edges of the form

$$\{v_0, v_1\}, \{v_1, v_2\}, ..., \{v_{m-1}, v_m\}$$

(also denoted by $v_0 \to v_1 \to v_2 \to ... \to v_m$). It is clear that an edge-sequence has the property that any two consecutive edges are either adjacent or identical; however, an arbitrary sequence of edges of G which has this property is not necessarily an edge-sequence (e.g. consider a star graph, and take its edges in any order). An edge-sequence trivially determines a sequence of vertices $v_0, v_1, ..., v_m$; we call v_0 the **initial vertex** and v_m the **final vertex** of the edge-sequence, and speak of an **edge-sequence from v_0 to v_m**. Note that if v_0 is any vertex, then the trivial edge-sequence which contains no edges is an edge-sequence from v_0 to v_0. The number of edges in an edge-sequence is called its **length**; for example, in Fig. 5.1, $v \to w \to x \to y \to z \to z \to y \to w$ is an edge-sequence of length seven from v to w.

The concept of edge-sequence is too general for our purposes, so we shall impose some further restrictions. An edge-sequence in which all the edges are distinct is called a **path**; if, in addition, the vertices $v_0, v_1, ..., v_m$ are distinct (except, possibly, $v_0 = v_m$), then the path is called a **chain**. A path or chain is **closed** if $v_0 = v_m$, and a closed chain containing at least one edge is called a **circuit**; note that in particular any loop or any pair of multiple edges forms a circuit.

REMARK. This is another instance of widely differing terminology by various authors. An edge-sequence appears in the literature as a walk, route, path or edge-progression; a path appears as a trail, semi-simple path, or chain; a chain as a path, arc, simple path or simple chain; a closed path as a cyclic path, re-entrant path or circuit; and a circuit as a cycle, circular path or simple circuit!

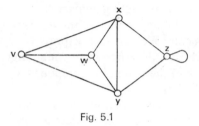

Fig. 5.1

In order to clarify the above concepts, let us again consider Fig. 5.1. We see that $v \to w \to x \to y \to z \to z \to x$ is a path, $v \to w \to x \to y \to z$ is a chain, $v \to w \to x \to y \to z \to x \to v$ is a closed path, and $v \to w \to x \to y \to v$ is a circuit. A circuit of length three (such as $v \to w \to x \to v$) is called a **triangle**. The extensions of all of these concepts to infinite graphs and digraphs are postponed until the relevant sections (§8 and §22).

We can now give an alternative, and possibly more useful, definition of a connected graph. A graph G is said to be **connected** if given any pair of vertices v, w of G, there is a chain from v to w. An arbitrary graph can be split up into disjoint connected sub-graphs called **(connected) components** by defining an equivalence relation on the vertex-set of G, two vertices being equivalent (or **connected**) if there is a chain from one to the other; we leave it to the reader to verify that the connectedness of vertices is in fact an equivalence relation, and that each of the required components can be obtained by taking the vertices in an equivalence class and the edges incident to them. Clearly a connected graph has only one component; a graph with more than one is called **disconnected.** We shall now prove that these definitions are consistent with the ones given in §3.

THEOREM 5A. *A graph is connected in the above sense if and only if it is connected in the sense of* §3.

Proof. ⇒ Let G be a graph which is connected in the above sense. If G is the union of two (disjoint) subgraphs, and v and w are two vertices, one from each subgraph, then any chain from v to w must contain an edge which is incident to a vertex of each subgraph; since no such edge exists, we have a contradiction.

⇐ Now suppose that G is connected in the sense of §3, and suppose that there is no chain connecting a given pair of vertices v and w; if we define connected components as above, then v and w will lie in different components. We can then express G as the union of two graphs, one of which is the component containing v and the other of which is the union of the remaining components; this establishes the required contradiction.//

Now that we know what connectedness means, it is natural to try to find out something about connected graphs. One direction of interest is to investigate bounds for the number of edges of a simple graph on n vertices with a given number of components. If such a graph is connected, it seems reasonable to expect that the graph has fewest edges when it has no circuits—such a graph is called a **tree**—and most edges when it is a complete graph; this would imply that the number of edges must lie between $n-1$ and $\frac{1}{2}n(n-1)$. We shall, in fact, prove a stronger theorem which includes this result as a special case.

THEOREM 5B. *Let G be a simple graph on n vertices; if G has k components, then the number m of edges of G satisfies*

$$n-k \leqq m \leqq \tfrac{1}{2}(n-k)(n-k+1).$$

Proof. To prove that $m \geqq n-k$, we use induction on the number of edges of G, the result being trivial if G is a null graph. If G contains as few edges as possible (say m_0), then the removal of any edge of G must increase the number of components by one, and the graph which remains will have n vertices, $k+1$ components, and m_0-1

edges. It follows from the induction hypothesis that $m_0 - 1 \geq n - (k+1)$, from which we immediately deduce that $m_0 \geq n-k$, as required.

To prove the upper bound, we can assume that each component of G is a complete graph. Suppose, then, that there are two components C_i and C_j with n_i and n_j vertices respectively, where $n_i \geq n_j > 1$. If we replace C_i and C_j by complete graphs on $n_i + 1$ and $n_j - 1$ vertices, then the total number of vertices remains unchanged, and the number of edges is increased by

$$\tfrac{1}{2}\{(n_i+1)n_i - n_i(n_i-1)\} - \tfrac{1}{2}\{n_j(n_j-1) - (n_j-1)(n_j-2)\} = n_i - n_j + 1,$$

which is positive. It follows that in order to attain the maximum number of edges, G must consist of a complete graph on $n-k+1$ vertices and $k-1$ isolated vertices; the result now follows immediately.//

COROLLARY 5C. *Any simple graph with n vertices and more than $\tfrac{1}{2}(n-1)(n-2)$ edges is connected.*//

Another approach used in the study of connected graphs is to ask the question, 'how connected is a connected graph?' One possible interpretation of this question is to ask how many edges must be removed from the graph in order to disconnect it. We conclude this section with a couple of definitions which are useful when discussing such a question; a **disconnecting set** of a connected graph G is a set of edges of G whose removal disconnects G; for example, in the graph of Fig. 5.2, the sets $\{e_1, e_2, e_5\}$ and $\{e_3, e_6, e_7, e_8\}$ are both

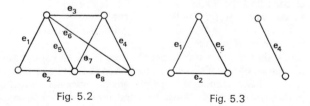

Fig. 5.2   Fig. 5.3

disconnecting sets of G, the disconnected graph left after removal of the second of these being shown in Fig. 5.3. We further define a **cutset** to be any disconnecting set, no proper subset of which is a disconnecting set; thus, in the example just given, only the second disconnecting set is actually a cutset. It is clear that the removal of the edges in a cutset always leaves a graph with exactly two components. If a cutset contains only one edge e, we shall call e an **isthmus** (see Fig. 5.4).

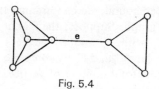

Fig. 5.4

These definitions can clearly be extended to disconnected graphs: if G is any graph, then a **disconnecting set** of G is a set of edges of G whose removal increases the number of components of G; a **cutset** of G is then simply a disconnecting set, no proper subset of which is a disconnecting set. Note that the removal of the edges in a cutset increases the number of components of G only by one.

As it turns out, there is a striking and unexpected similarity between the properties of circuits and those of cutsets; the reader may recognize this if he looks at such exercises as *5f*, *5i*, *5j*, *6h* and *6i*. The reason for this will be revealed in Chapter IX, when everything will suddenly become clear!

Exercises

(*5a*)　In the Petersen graph, find (*i*) an edge-sequence of length four; (*ii*) circuits of lengths five, six, eight and nine; (*iii*) cutsets containing three, four and five edges.

(*5b*)　The **girth** of a graph is defined to be the length of its shortest circuit; find the girths of (*i*) K_n; (*ii*) $K_{m,n}$; (*iii*) C_n; (*iv*) W_n; (*v*) the Platonic graphs; (*vi*) the Petersen graph.

(*5c*)　Show that an edge of a connected graph is an isthmus if and only if it is contained in no circuit.

(*5d*)　Show that if G is a simple graph, then (*i*) G and $\bar{G}$ cannot both be disconnected; (*ii*) if G is connected, then so is its line graph $L(G)$.

(*5e*)　Let G be a connected graph with vertex-set $\{v_1, ..., v_n\}$ and adjacency matrix A. Show that the number of edge-sequences of length k from v_i to v_j is the ij-th entry of the matrix A^k. Show also that if G is a simple graph and c denotes the number of triangles in G, then $6c = \text{trace } (A^3)$. (The trace of a square matrix is the sum of the entries along its main diagonal.)

(*5f*)　Show that if two distinct circuits of a graph G each contain an edge e, then there exists a circuit of G which does not contain e; show also that a similar result holds if the word 'circuit' is replaced throughout by 'cutset'.

(5g)   Show that a graph is bipartite if and only if all of its circuits have even length. Can you find an analogous result for cutsets?

(5h)   Let G be a simple graph in which $\rho(v) \geq r$ ($r \geq 2$) for every vertex v; show that G contains a circuit of length at least $r + 1$.

(5i)   Let C be a circuit and C^* a cutset of a connected graph; show that the number of edges common to C and C^* is even. Show also that if a set S of edges has an even number of edges in common with every cutset, then S may be split up into circuits, no two of which have any edges in common.

(5j)   A set E of edges of a graph is said to be **independent** if E contains no circuit; show that (i) every subset of an independent set is independent; (ii) if I and J are independent sets containing k and $k + 1$ edges respectively, then there exists an edge e which is in J but not in I, with the property that $I \cup \{e\}$ is an independent set. Show also that (i) and (ii) still hold if we replace the word 'circuit' by 'cutset'.

(*5k) Let G be a simple graph on $2n$ vertices which contains no triangles. Use induction on n to show that G has at most n^2 edges, and give an example of a graph in which this upper bound is achieved. (This result is known as **'Turán's extremal theorem'**.

(5l)   In a connected graph, let $d(v, w)$ (the **distance** between the vertices v and w) denote the length of the shortest chain from v to w. Show that the function d defines a metric on the vertex-set, satisfying (i) for any v and w, $d(v, w)$ is an integer; (ii) if $d(v, w) \geq 2$, then there exists a vertex z such that $d(v, z) + d(z, w) = d(v, w)$. Show, conversely, that any finite metric space in which (i) and (ii) are satisfied is isomorphic to the metric space of some graph.   ‹metric space›

(*5m) The **diameter** δ of a connected graph G with n vertices is the maximum possible distance between any two vertices; a **centre** of G is a vertex v with the property that the maximum possible distance between v and any other vertex is as small as possible, this distance being called the **radius** r (thus

$$r = \min_{v} \max_{w} d(v, w)).$$

Give an example of a graph with more than one centre. Find the diameters and radii of K_n, C_n, W_n and the Petersen graph. Show also that $r \leq \delta \leq 2r$, and that $\log(n\rho - n + 1) \leq (r + 1) \log \rho$ where ρ is the largest vertex-degree.

(*5n) How would you define edge-sequences, chains, connectedness, and cutsets for a k-graph?

§6. EULERIAN GRAPHS

A connected graph G is called **Eulerian** if there exists a closed path which includes every edge of G; such a path is then called an

Eulerian path. Note that the definition requires each edge to be traversed once and once only. *G* is **semi-Eulerian** if we remove the restriction that the path must be closed; thus every Eulerian graph is semi-Eulerian. Figs. 6.1, 6.2 and 6.3 show graphs which are

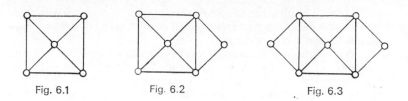

Fig. 6.1   Fig. 6.2   Fig. 6.3

non-Eulerian, semi-Eulerian and Eulerian respectively. Note that the assumption that *G* is connected is merely a technicality introduced in order to avoid the trivial case of a graph containing several isolated vertices.

Problems on Eulerian graphs frequently appear in books on recreational mathematics—a typical problem might ask whether a given diagram can be drawn without lifting one's pencil from the paper and without repeating any lines. The name 'Eulerian' arises from the fact that Euler was the first person to solve the famous Königsberg bridge problem which asked, in effect, whether the graph in Fig. 6.4 has an Eulerian path (it hasn't!).

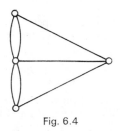

Fig. 6.4

One question which immediately arises is 'can one find necessary and sufficient conditions for a graph to be Eulerian?' Before providing a complete answer to this question in theorem 6B, we prove a simple lemma.

LEMMA 6A. *If G is a graph in which the degree of every vertex is at least two, then G contains a circuit.*

Proof. If G contains any loops or multiple edges, the result is trivial; we can therefore suppose that G is a simple graph. Let v be any vertex of G; we can construct an edge-sequence $v \to v_1 \to v_2 \dots$ inductively by choosing v_1 to be any vertex adjacent to v, and for $i \geqq 1$ choosing v_{i+1} to be any vertex adjacent to v_i except v_{i-1} (the existence of such a vertex v_{i+1} being guaranteed by our hypothesis). Since G has only finitely many vertices, we must eventually choose a vertex which has been chosen before; if v_k is the first such vertex, then that part of the edge-sequence which lies between the two occurrences of v_k is the required circuit.//

THEOREM 6B. *A connected graph G is Eulerian if and only if the degree of every vertex of G is even.*

Proof. $\Rightarrow$ Suppose that P is an Eulerian path of G. Whenever P passes through any vertex, there is a contribution of two towards the degree of that vertex; since every edge occurs exactly once in P, every vertex must have even degree.

$\Leftarrow$ The proof is by induction on the number of edges of G. Since G is connected, every vertex has degree at least two, and so by the above lemma, G contains a circuit C. If C contains every edge of G,

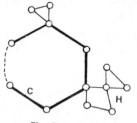

Fig. 6.5

the proof is complete; if not, we remove from G the edges of C to form a new (possibly disconnected) graph H which has fewer edges than G and in which every vertex still has even degree. By the induction hypothesis, each component of H has an Eulerian path. Since each component of H has at least one vertex in common with C, by connectedness, we obtain the required Eulerian path of G by following the edges of C until a non-isolated vertex of H is reached, tracing the Eulerian path of the component of H which contains that vertex, and then continuing along the edges of C until we reach a vertex belonging to another component of H, and so on; the whole process terminates when we get back to the initial vertex (see Fig. 6.5).//

The proof just given can be easily modified to prove the following two results. We leave the details to the reader.

COROLLARY 6C. *A connected graph is Eulerian if and only if its edge-family can be split up into disjoint circuits.*//

COROLLARY 6D. *A connected graph is semi-Eulerian if and only if there are not more than two vertices of odd degree.*//

We remark that if a semi-Eulerian graph has exactly two vertices of odd degree, then any semi-Eulerian path (in the obvious sense) must have one of them as initial vertex and the other as final vertex. By the handshaking lemma, a graph cannot have exactly one vertex of odd degree.

★ To conclude our discussion of Eulerian graphs, we now give an algorithm for constructing an Eulerian path in a given Eulerian graph. The method is known as **Fleury's algorithm.**

THEOREM 6E. *Let G be an Eulerian graph; then the following construction is always possible, and produces an Eulerian path of G. Start at any vertex u and traverse the edges in an arbitrary manner, subject only to the following rules:*

(i) erase the edges as they are traversed, and if any isolated vertices result erase them too; (ii) at each stage, use an isthmus only if there is no alternative.

Proof. We shall show first that at each stage the construction may be carried out. Suppose that we have just reached a vertex v; then if $v \neq u$, the subgraph H which still remains is connected and contains only two vertices of odd degree—namely, u and v. By corollary 6D, H contains a semi-Eulerian path P from v to u. Since the removal of the first edge of P does not disconnect H, it follows that at each stage the construction is possible. If $v = u$, the proof is almost identical, as long as there are still edges incident with u.

It remains only to show that the construction always yields a complete Eulerian path. But this is clear, since there can be no edges of G remaining untraversed when the last edge incident to u is used (since otherwise the removal of some earlier edge adjacent to one of these edges would have disconnected the graph, contradicting (ii)).//★

Exercises

(6a) For which values of m and n are the following graphs Eulerian;
(*i*) $K_{m,n}$, (*ii*) K_n, (*iii*) W_n? Are any of the Platonic graphs Eulerian?
If so, find an Eulerian path.

(6b) If G is connected and has k (>0) vertices of odd degree, show that
the minimum number of paths which together include every edge
of G and no two of which have an edge in common, is $\frac{1}{2}k$; deduce
corollary 6D as a special case. What happens if k is odd?

(6c) Use Fleury's algorithm to produce an Eulerian path for the graph
shown in Fig. 6.6.

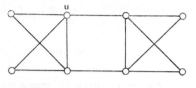

Fig. 6.6

(6d) Is it possible for a knight to travel round an 8×8 chessboard in such
a way that every possible move occurs exactly once, (a move
between two squares being said to 'occur' if it is traversed in either
direction)? Repeat the question for a king and a rook. How would
your answers be changed if the chessboard were a 7×7 one?
Describe all your answers in graph-theoretic terms.

(6e) Show that the line graph of an Eulerian simple graph is Eulerian.
Given that the line graph of a simple graph G is Eulerian, can we
deduce that G is Eulerian?

(6f) An Eulerian graph is **arbitrarily-traceable** from a vertex v if, whenever we start from v and traverse the graph in an arbitrary way,
never using any edge twice, we eventually obtain an Eulerian path.
Show that the graph shown in Fig. 6.7 is arbitrarily-traceable, and
give an example of an Eulerian graph which is not. Can you find
any Eulerian graphs which are arbitrarily-traceable from two
distinct vertices? Can you suggest why an arbitrarily-traceable
graph might be used as a suitable layout for an exhibition?

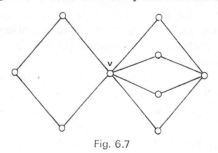

Fig. 6.7

(*6g) Show that K_5 has 264 Eulerian paths. Given a set of fifteen dominoes from 0–0 to 4–4, in how many distinct ways can they be laid out in a circuit with corresponding numbers lying next to each other?

(*6h) Let V be the vector space associated (in the sense of exercise 2j) with a graph G. Use corollary 6c to show that if C and D are circuits of G, then their vector sum $C \oplus D$ may be written as an edge-disjoint union of circuits; deduce that the set of such unions of circuits of G forms a subspace W of V (called the **circuit subspace** of G).

(*6i) Using the notation of the previous exercise, show analogously that the set of edge-disjoint unions of cutsets of G forms a subspace $\tilde{W}$ of V (called the **cutset subspace** of V). Find the dimensions of W and $\tilde{W}$, and deduce that V may be expressed as the direct sum (in the sense of linear algebra) of W and $\tilde{W}$. (See also exercise 9k.)

§7. HAMILTONIAN GRAPHS

In the previous section we discussed the problem of whether there exists a closed path which includes every edge of a given connected graph G. A similar problem we can consider is whether there exists

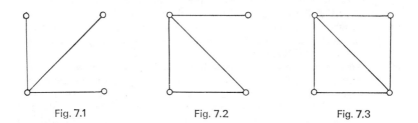

Fig. 7.1   Fig. 7.2   Fig. 7.3

a closed path which passes exactly once through each vertex of G. It is clear that such a path must be a circuit (excluding the trivial case in which G is the graph N_1); if such a circuit exists, it is called a **Hamiltonian circuit**, and G is called a **Hamiltonian graph**. A graph which contains a chain which passes through every vertex is called **semi-Hamiltonian**; note that every Hamiltonian graph is semi-Hamiltonian. Figs. 7.1, 7.2 and 7.3 show graphs which are non-Hamiltonian, semi-Hamiltonian and Hamiltonian respectively.

The name 'Hamiltonian circuit' arises from the fact that Sir William Hamilton investigated the existence of such circuits in the dodecahedral graph; such a circuit is shown in Fig. 7.4, heavy lines

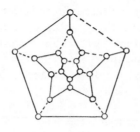

Fig. 7.4

denoting its edges. It is interesting to note that Hamilton's paper was one of the earliest papers involving group theory (see exercise 7e).

In theorem 6B we obtained a necessary and sufficient condition for a connected graph to be Eulerian, and it is perhaps reasonable to expect that we can obtain a similar characterization for Hamiltonian graphs. As it happens, the finding of such a characterization is one of the major unsolved problems of graph theory! In fact, little is known in general about Hamiltonian graphs. Most existing theorems have the form, 'if G has enough edges, then G is Hamiltonian'; probably the most celebrated of these is the following one, due to G. A. Dirac, and known, reasonably enough, as **Dirac's theorem.**

THEOREM 7A (Dirac 1952). *If G is a simple graph with n vertices, and $\rho(v) \geqq \frac{1}{2}n$ for every vertex v, then G is Hamiltonian.*

Remark. There are several proofs of this well-known theorem; the proof given here is due to D. J. Newman.

Proof. In order to prove this theorem, we introduce k new vertices, each of which is joined to every vertex of G; we shall suppose that k is the smallest number of such vertices necessary to make the resulting graph G' Hamiltonian. We assume that $k > 0$, and derive a contradiction.

Let $v \rightarrow p \rightarrow w \rightarrow \ldots \rightarrow v$ be a Hamiltonian circuit of G', where v and w are vertices of G, and p is one of the new vertices. Then w cannot be adjacent to v, since otherwise we could omit p, contradicting the minimality of k. Moreover, a vertex (w', say) which is adjacent to w cannot immediately follow a vertex v' which is adjacent to v, since we could then replace $v \rightarrow p \rightarrow w \rightarrow \ldots \rightarrow v' \rightarrow w' \rightarrow \ldots \rightarrow v$ by $v \rightarrow v' \rightarrow \ldots \rightarrow w \rightarrow w' \rightarrow \ldots \rightarrow v$, by reversing the part between w and v'. Hence the number of vertices of G' which are not adjacent to w is at least as great as the number of vertices adjacent to v (i.e. at least $\frac{1}{2}n+k$); clearly the number of vertices of G' which are adjacent to w is also at least $\frac{1}{2}n+k$. Since no vertex of G' can be both adjacent and not-adjacent to w, it follows that the total number, $n+k$, of vertices of G' is not less than $n+2k$, giving the required contradiction.//

Exercises

(7a) For which values of m and n are the following graphs Hamiltonian; (i) $K_{m,n}$; (ii) K_n; (iii) W_n? Describe a Hamiltonian circuit in each case. Show also that the Platonic graphs are all Hamiltonian, and find a Hamiltonian circuit for each.

(7b) Show that the Petersen graph is non-Hamiltonian; is it semi-Hamiltonian?

(7c) Give an example of a graph which is Eulerian but not Hamiltonian, and one which is Hamiltonian but not Eulerian. What can you say about graphs which are both Eulerian and Hamiltonian?

(7d) Is it possible for a knight to visit all the squares of an 8×8 chessboard exactly once, and then return to its starting-point? Repeat the question for a king and a rook. How would your answers be changed if the chessboard were a 7×7 one? Describe all your answers in a graph-theoretic terms.

(7e) Let H be the group generated by the two elements l and r subject to the relations $r^5 = 1$, $lr^2l = rlr$ and $lr^3l = r^2$ (where 1 denotes the identity element of H). Show that $(lrlr^3l^3r)^2 = 1$. What is the connexion between this problem and the problem of finding a Hamiltonian circuit in the dodecahedral graph?

(7f) Give a counter-example to show that in the statement of Dirac's theorem, the condition '$\rho(v) \geqq \frac{1}{2}n$' cannot be replaced by '$\rho(v) \geqq \frac{1}{2}n-1$'.

(7g) Show that if a simple graph G has n ($\geqq 3$) vertices, and for every pair v, w of non-adjacent vertices $\rho(v)+\rho(w) \geqq n$, then G is Hamiltonian. Deduce Dirac's theorem.

(*7h) Show that the line graphs of both Eulerian and Hamiltonian simple graphs are Hamiltonian. Given that the line graph of a simple graph G is Hamiltonian, can we deduce that G is either Eulerian or Hamiltonian?

(*7i) Find n Hamiltonian circuits in K_{2n+1}, with the property that no two of them have an edge in common. Nine gourmets visit their favourite restaurant every night during a conference; if their table is circular and no two of them sit next to each other more than once, what can you say about the length of the conference? Repeat the question for a bipartite graph of the form $K_{n,n}$, and find a similar 'practical' application in this case.

§8. INFINITE GRAPHS

In this section we show how some of the definitions given in previous sections can be extended to infinite graphs. As the reader will recall, an **infinite graph** G is a pair $(V(G), E(G))$, where $V(G)$ is an infinite set of elements called **vertices,** and $E(G)$ is an infinite family of unordered pairs of elements of $V(G)$ called **edges.** If $V(G)$ and $E(G)$ are both countably infinite, then G is said to be a **countable graph.** Note that we have excluded from these definitions the possibility of $V(G)$ being infinite but $E(G)$ finite (such objects being merely finite graphs together with infinitely many isolated vertices) or of $E(G)$ being infinite but $V(G)$ finite (such objects being essentially finite graphs but with infinitely many loops or multiple edges).

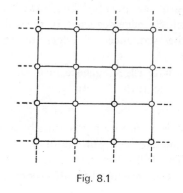

Fig. 8.1

Several of the definitions given in Chapter II (for example, 'adjacent', 'incident', 'isomorphic', 'subgraph', 'union', 'connected', 'component') generalize immediately to infinite graphs. The **degree** of a vertex v of an infinite graph is defined to be the cardinality of the set of edges incident to v, and may be finite or infinite; an infinite graph, all of whose vertices have finite degree is called **locally-finite,** a well-known example being the infinite square lattice, part of which is shown in Fig. 8.1. We can similarly define a **locally-countable** infinite graph to be one, all of whose vertices have countable degree. With these definitions, we now prove the following simple, but fundamental, result.

THEOREM 8A. *Every connected locally-countable infinite graph is countable.*

Proof. Let v be any vertex of such an infinite graph, and let A_1 be the set of vertices adjacent to v, A_2 the set of all vertices adjacent to a vertex of A_1, and so on. By hypothesis, A_1 is countable, and hence so are A_2, A_3 ... (using the fact that the union of a countable collection of countable sets is countable); hence $\{v\}$, $A_1, A_2, ...$ is a sequence of sets whose union is countable. Moreover, this sequence contains every vertex of the infinite graph, by connectedness, and the result follows.//

COROLLARY 8B. *Every connected locally-finite infinite graph is countable.*//

We can also extend to an infinite graph G the concept of an edge-sequence, there being essentially three different types:

(*i*) a **finite edge-sequence** in G is defined exactly as in §**5**;

(*ii*) a **one-way infinite edge-sequence** in G (with **initial vertex** v_0) is an infinite sequence of edges of the form $\{v_0, v_1\}$, $\{v_1, v_2\}$, ...;

(*iii*) a **two-way infinite edge-sequence** in G is an infinite sequence of edges of the form ..., $\{v_{-2}, v_{-1}\}$, $\{v_{-1}, v_0\}$, $\{v_0, v_1\}$, $\{v_1, v_2\}$, ...

One-way and two-way infinite paths and chains are defined in the obvious way, as are such terms as the length of a path and the distance between vertices. The following result, known as **König's lemma,** tells us that infinite chains are not difficult to come by:

THEOREM 8C (König 1936). *Let G be a connected locally-finite infinite graph; then for any vertex v of G, there exists a one-way infinite chain with initial vertex v.*

Proof. If z is any vertex of G other than v, then there is a non-trivial chain from v to z; it follows that there are infinitely many chains in G with initial vertex v. Since the degree of v is finite, there must be infinitely many of these chains which start with the same edge. If $\{v, v_1\}$ is such an edge, then we can repeat this procedure for the vertex v_1 and thus obtain a new vertex v_2 and a corresponding edge $\{v_1, v_2\}$. By carrying on in this way, we obtain the one-way infinite chain $v \to v_1 \to v_2 ... .//$

The importance of König's lemma is that it allows us to deduce results about infinite graphs from the corresponding results for finite graphs; the following theorem (in which we anticipate some of the definitions and results of Chapter V) may be regarded as a typical example:

THEOREM 8D. *Let G be a countable graph, every finite subgraph of which is planar; then G is planar.*

Proof. Since G is countable, its vertices may be enumerated as v_1, v_2, v_3, ... ; we now construct a strictly increasing sequence $G_1 \subset G_2 \subset G_3 \subset$ ... of subgraphs of G, by taking G_k to be the subgraph whose vertices are precisely $v_1, ..., v_k$ and whose edges are those edges of G which join two of these vertices. Then, assuming the result that G_i can be embedded in the plane in only a finite number ($m(i)$, say) of topologically distinct ways, we can construct another infinite graph H whose vertices w_{ij} ($i \geq 1$, $1 \leq j \leq m(i)$) correspond to the various embeddings of the graphs $\{G_i\}$, and whose edges join those vertices w_{ij} and w_{kl} for which $k = i+1$ and the plane embedding corresponding to w_{kl} 'extends' (in an obvious sense) the embedding corresponding to w_{ij}. Since H is clearly connected and locally-finite, it follows from König's lemma that H contains a one-way infinite chain; since G is countable, this infinite chain gives the required plane embedding of the whole of G.//

It is worth pointing out that if we assume further axioms of set theory (in particular, the uncountable version of the axiom of choice), then various results such as the one just proved can be extended to infinite graphs which are not necessarily countable.

★ We conclude this digression on infinite graphs with a brief discussion on infinite Eulerian graphs. It seems natural to say that a connected infinite graph G is **Eulerian** if there exists a two-way infinite path which includes every edge of G; such an infinite path is then called a (two-way) **Eulerian path.** We can further say that G is **semi-Eulerian** if there exists an infinite path (one-way or two-way) which includes every edge of G. Note that these definitions require G to be countable; the following theorems give further conditions which are necessary for an infinite graph to be Eulerian or semi-Eulerian.

THEOREM 8E. *Let G be a connected countable graph which is Eulerian; then* (i) *G has no vertices of odd degree;* (ii) *for every finite subgraph H of G, the infinite graph $\overline{H}$ (obtained by removing from G the edges of H) has at most two infinite connected components;* (iii) *if, in addition, every vertex of H has even degree, then $\overline{H}$ has exactly one infinite connected component.*

C

Proof. (*i*) Suppose that P is an Eulerian path; then by the argument given in the proof of theorem 6B, every vertex of G must have either even or infinite degree.

(*ii*) Let P be split up into three subpaths P_-, P_0, and P_+, in such a way that P_0 is a finite path containing every edge of H (and possibly other edges as well), and P_- and P_+ are both one-way infinite paths. Then the infinite graph K formed by the edges of P_- and P_+ (and the vertices incident to them) has at most two infinite components; since $\bar{H}$ is obtained by adding only a finite set of edges to K, the result follows.

(*iii*) Let the initial and final vertices of P_0 be v and w; we wish to show that v and w are connected in $\bar{H}$. If $v = w$, this is obvious; if not, then the result follows on applying corollary 6D to the graph obtained by removing from P_0 the edges of H, this graph having exactly two vertices (v and w) of odd degree, by hypothesis.//

We can obtain corresponding necessary conditions for semi-Eulerian infinite graphs; the proof is similar to (but easier than) the proof of theorem 8E, and will be left as an exercise.

THEOREM 8F. *Let G be a connected countable graph which is semi-Eulerian but not Eulerian; then (i) G has either at most one vertex of odd degree or at least one of infinite degree; (ii) for every finite subgraph H of G, the infinite graph $\bar{H}$ (obtained as before) has exactly one infinite component.//*

It turns out that the conditions given in the previous two theorems are not only necessary but also sufficient. We state this result formally in the following theorem; its proof lies beyond the scope of this book, but may be found in Ore.[4]

THEOREM 8G. *Let G be a connected countable graph; then G is Eulerian if and only if the conditions (i), (ii) and (iii) of theorem 8E are satisfied. Moreover, G is semi-Eulerian if and only if either these conditions or conditions (i) and (ii) of theorem 8F are satisfied.//*★

Exercises

(*8a*) Find an analogue of theorem 8A for infinite graphs whose vertex-degrees are higher cardinals?

(*8b*) Prove theorem 8F.

(*8c*) Find a generalization to infinite graphs of corollary 6C.

(*8d*)   Show by a counter-example that the conclusion of König's lemma is false if we omit the condition that the infinite graph must be locally-finite.

(*8e*)   Let S_2 denote the infinite square lattice; show (by finding an explicit Eulerian path) that S_2 is Eulerian, and verify in detail that S_2 satisfies the conditions of theorem 8E.

(*8f*)   Are the infinite triangular lattice T_2 and the infinite hexagonal lattice H_2 Eulerian? If so, find an explicit Eulerian path for each.

(*8g*)   Show that S_2 contains both one-way and two-way infinite chains which pass exactly once through each vertex; do analogous results hold for T_2 and H_2?

4   Trees

We are all familiar with the idea of a family tree; in this chapter, we shall be studying trees in general, with special reference to spanning trees in a connected graph and (in §10) to Cayley's celebrated result on the enumeration of labelled trees. The chapter concludes with a section on some applications of graph theory.

§9. ELEMENTARY PROPERTIES OF TREES

A **forest** is defined to be a graph which contains no circuits, and a connected forest is called a **tree**; for example, Fig. 9.1 shows a forest

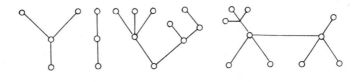

Fig. 9.1

with four components, each of which is a tree.† Note that trees and forests are by definition simple graphs.

In many ways a tree is the simplest non-trivial type of graph; as we shall see in theorem 9A, it has several 'nice' properties such as the fact that any two vertices are connected by a unique chain. In trying to prove a general result or test a general conjecture in graph theory, it is sometimes convenient to start by trying to prove the corresponding result for a tree; in fact there are several conjectures which have not been proved for arbitrary graphs but which are known to be true for trees.

---

† The last tree in Fig. 9.1 is particularly well-known for its bark.

The following theorem lists some simple properties of trees:

THEOREM 9A. *Let T be a graph with n vertices. Then the following statements are equivalent:* (*i*) T *is a tree;* (*ii*) T *contains no circuits, and has $n-1$ edges;* (*iii*) T *is connected, and has $n-1$ edges;* (*iv*) T *is connected, and every edge is an isthmus;* (*v*) *any two vertices of T are connected by exactly one chain;* (*vi*) T *contains no circuits, but the addition of any new edge creates exactly one circuit.*

Proof. If $n = 1$, all six results are trivial. We shall therefore assume that $n \geq 2$.

(*i*) $\Rightarrow$ (*ii*). T contains no circuits by definition; hence, by exercise 5c, the removal of any edge disconnects T into two graphs, each of which is a tree. It follows by induction that the number of edges in each of these two trees is one fewer than the number of vertices, from which we deduce that the total number of edges of T is $n-1$.

(*ii*) $\Rightarrow$ (*iii*). If T is disconnected, then each component of T is a connected graph with no circuits and hence, by the previous part, the number of vertices in each component exceeds by one the number of edges. It follows that the total number of vertices of T exceeds the total number of edges by at least two, contradicting the fact that T has $n-1$ edges.

(*iii*) $\Rightarrow$ (*iv*). The removal of any edge results in a graph with n vertices and $n-2$ edges, which must be disconnected by theorem 5B.

(*iv*) $\Rightarrow$ (*v*). Since T is connected, each pair of vertices is connected by at least one chain. If a given pair of vertices is connected by two chains, then they enclose a circuit, which contradicts the fact that every edge is an isthmus (by exercise 5c).

(*v*) $\Rightarrow$ (*vi*). If T contained a circuit, then any two vertices in the circuit would be connected by at least two chains. If an edge e is added to T, then, since the vertices incident to e are already connected in T, a circuit will be created; the fact that only one circuit is obtained follows from exercise 5f.

(*vi*) $\Rightarrow$ (*i*). Suppose that T is disconnected; if to T we add any edge which joins a vertex of one component to a vertex in another, then no circuit will be created.//

COROLLARY 9B. *Let G be a forest with n vertices and k components; then G has $n-k$ edges.*

Proof. Apply the above statement (*ii*) to each component of G.//

Note that by the handshaking lemma, the sum of the degrees of all the n vertices of a tree is equal to twice the number of edges $(=2n-2)$; it follows that if $n \geq 2$, a tree on n vertices always contains at least two terminal vertices.

Given any connected graph G, we can choose a circuit and remove one of its edges, the resulting graph remaining connected (by exercise 5c). We repeat this procedure with one of the remaining circuits, continuing until there are no circuits left. The graph which remains will be a tree which connects all the vertices of G; it is called a **spanning tree** of G. An example of a graph and one of its spanning trees appears in Fig. 9.2 and 9.3.

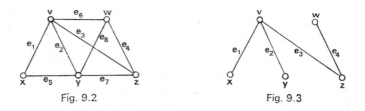

Fig. 9.2   Fig. 9.3

More generally, if G now denotes an arbitrary graph with n vertices, m edges and k components, we can carry out the above procedure on each component of G, the result being called a **spanning forest** (or **skeleton**). The number of edges removed in the process is called the **circuit rank** (or **cyclomatic number**) of G, and is denoted by $\gamma(G)$; note that $\gamma(G) = m-n+k$, which is a nonnegative integer by theorem 5B. We see that the circuit rank gives a measure of the connectedness of a graph (in a sense to be made precise in exercise 9k)—the circuit rank of a tree is zero and of a circuit graph is one. It is convenient also to define the **cutset rank** (or **component rank**) of G to be the number of edges in a spanning forest; it is denoted by $\kappa(G)$ and is equal to $n-k$.

Before proceeding, we shall prove a couple of simple results concerning spanning forests; in this theorem, the **complement** of a spanning forest T of a (not necessarily simple) graph G is simply the graph obtained from G by removing the edges of T.

THEOREM 9C. *If T is any spanning forest of a graph G, then (i) every cutset of G has an edge in common with T, and (ii) every circuit of G has an edge in common with the complement of T.*

Proof. (*i*) Let C^* be a cutset of G, the removal of which splits one of the components of G into two subgraphs H and K. Then since T is a spanning forest, T must contain an edge joining a vertex of H to a vertex of K; this edge is the required edge.

(*ii*) Let C be a circuit of G which has no edge in common with the complement of T; then C must be contained in T, which is a contradiction.//

Closely linked with the idea of a spanning forest T of a graph G, is the concept of the fundamental system of circuits associated with T, formed as follows: if we add to T any edge of G not contained in T, then by statement (*vi*) of theorem 9A we get a unique circuit. The set of all circuits formed in this way (i.e. by adding separately each edge of G not contained in T) is called the **fundamental system of circuits associated with T.** Sometimes we are not interested in the particular spanning forest chosen, and refer simply to a **fundamental system of circuits of G.** In any case, it is clear that the circuits in a given fundamental system must be distinct, and that the number of such

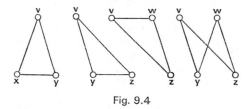

Fig. 9.4

circuits must be equal to the circuit rank of G. Fig. 9.4 shows the fundamental system of circuits of the graph shown in Fig. 9.2 associated with the spanning tree of Fig. 9.3.

In the light of our remarks at the end of §5, we may hope to be able to define a fundamental system of cutsets of a graph G associated with a given spanning forest T; we shall now show that this is indeed the case. By statement (*iv*) of theorem 9A, the removal of any edge of T divides the set of vertices of T into two disjoint sets V_1 and V_2. The set of all edges of G joining a vertex of V_1 with one of V_2 is a cutset of G, and the set of all cutsets obtained in this way (i.e. by removing separately each edge of T) is called the **fundamental system of cutsets** associated with T. It is clear that the cutsets in a given fundamental system must be distinct, and that the number of such cutsets must be equal to the cutset rank of G. The fundamental system of cutsets of the graph in Fig. 9.2 associated with the spanning tree of Fig. 9.3 is $\{e_1, e_5\}$, $\{e_2, e_5, e_7, e_8\}$, $\{e_3, e_6, e_7, e_8\}$ and $\{e_4, e_6, e_8\}$.

Exercises

($9a$) Show that there are exactly six non-isomorphic trees on six vertices, and eleven on seven vertices.

($9b$) Show that every tree is a bipartite graph; which trees are complete bipartite graphs?

($9c$) Show that the graphs associated (in the sense of exercise $1b$) with the saturated hydrocarbons (C_nH_{2n+2}) and the alcohols ($C_nH_{2n+1}OH$) are in fact trees.

($9d$) Calculate the circuit and cutset ranks of (i) K_n; (ii) $K_{m,n}$; (iii) N_n; (iv) W_n; (v) the Platonic graphs; (vi) the Petersen graph; (vii) any connected graph on n vertices which is regular of degree r.

($9e$) Find a spanning tree and the corresponding fundamental systems of circuits and cutsets of (i) K_5; (ii) $K_{3,3}$; (iii) W_5; (iv) C_6; (v) the Platonic graphs; (vi) the Petersen graph.

($9f$) Show that every tree has either one or two centres.

($9g$) Let T_1 and T_2 be spanning trees of a connected graph G; show that if e is any edge of T_1, then there exists an edge f of T_2 with the property that $(T_1 - \{e\}) \cup \{f\}$ (the graph obtained from T_1 on replacing e by f) is also a spanning tree. Show also that T_1 can be 'transformed' into T_2 by replacing the edges of T_1 one at a time by edges of T_2 in such a way that at each stage we obtain a spanning tree.

($9h$) Show that if C^* is a set of edges of a graph G with the property that C^* has an edge in common with any spanning forest of G, then C^* contains a cutset; obtain a corresponding result for circuits.

(*$9i$) Let A be the incidence matrix of a tree on n vertices. Show that any $n-1$ columns of A are linearly independent over the field of integers modulo 2.

(*$9j$) Show that if H and K are subgraphs of a graph G, and $H \cup K$, $H \cap K$ are defined in the natural way, then the cutset rank κ satisfies: (i) $0 \leqq \kappa(H) \leqq m(H)$ (the number of edges of H); (ii) if H is a subgraph of K, then $\kappa(H) \leqq \kappa(K)$; ($iii$) $\kappa(H \cup K) + \kappa(H \cap K) \leqq \kappa(H) + \kappa(K)$.

(*$9k$) Let V be the vector space associated with a simple connected graph G, and let T be a spanning tree of G. Show that the fundamental system of circuits associated with T forms a basis for the circuit subspace W of G, and obtain a corresponding result for the cutset subspace $\tilde{W}$; deduce that the dimensions of W and $\tilde{W}$ are $\gamma(G)$ and $\kappa(G)$ respectively.

§10. THE ENUMERATION OF TREES

The subject of graph enumeration is concerned with the problem of finding out how many non-isomorphic graphs there are which possess a given property. The subject was probably initiated in the 1850s by Cayley, who was trying to count the number of saturated hydro-carbons C_nH_{2n+2} containing a given number of carbon atoms; as he realized, and as the reader saw in exercise 9c, this problem can be expressed as the problem of counting the number of trees in which the degree of every vertex is either four or one.

Many of the standard problems of graph enumeration have been solved. For example, it is possible to calculate the number of graphs, digraphs, connected graphs, trees and Eulerian graphs, containing a given number of vertices and edges; the corresponding results for planar graphs and Hamiltonian graphs have, however, not yet been obtained. Most of the known results can be obtained by using a fundamental enumeration theorem due to Pólya, a good account of which may be found in Liu[6]; unfortunately, in almost every case it is impossible to express these results by means of simple formulae. For tables of some known results, the reader is referred to the appendix.

This section is devoted to the proof of a famous result, usually attributed to Cayley, on the number of labelled trees with a given number of vertices. The reader has already met labelled graphs at the end of §2, a labelled graph on n vertices being essentially a graph in which the vertices are 'labelled' with the integers from 1 to n. More precisely, we define a **labelling** of a graph G on n vertices to be a one-one mapping from the vertex-set of G onto the set $\{1, ..., n\}$; a **labelled graph** is then a pair (G, ϕ), where G is a graph and ϕ is a labelling of G. We shall frequently refer to the integers $1, ..., n$ as the **labels** of G, and denote the vertices of G by $v_1, ..., v_n$. Further-more, we shall say that two labelled graphs (G_1, ϕ_1) and (G_2, ϕ_2) are **isomorphic** if there exists an isomorphism between G_1 and G_2 which preserves the labelling of the vertices.

In order to clarify these definitions, let us consider Fig. 10.1, which shows various ways of labelling a tree with four vertices. On closer inspection, we see that the second labelled tree is simply the reverse of the first one, and it follows that these two labelled trees must be isomorphic; on the other hand, neither of them is isomorphic to the third labelled tree (as can be seen by looking at the degree of the vertex v_3). It follows that the total number of ways of labelling this particular tree must be $\frac{1}{2}(4!) = 12$, since the reverse of any labelling does not result in a new one. Similarly, the total number of ways of labelling the tree shown in Fig. 10.2 must be four, since the

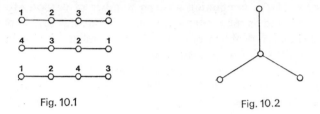

Fig. 10.1   Fig. 10.2

central vertex may be labelled in four different ways, and each one determines the labelling. It follows that the total number of (non-isomorphic) labelled trees on four vertices is sixteen; we now prove **Cayley's theorem** which generalizes this result to labelled trees with n vertices.

THEOREM 10A (Cayley 1889). *There are n^{n-2} distinct labelled trees on n vertices.*

Remark. The proof we are about to give is due to Clarke; for several other proofs, see Moon.[7]

Proof. Let $T(n, k)$ denote the number of labelled trees on n vertices in which a given vertex (v, say) has degree k. We shall derive an expression for $T(n, k)$, and the result will then follow on summing from $k = 1$ to $k = n-1$.

Let A be any labelled tree in which $\rho(v) = k-1$. The removal from A of any edge $\{w, z\}$ which is not incident to v leaves two subtrees, one of which contains v and either w or z (let us say, w), and the other of which contains z. If we now join the vertices v and z, we obtain a labelled tree B in which $\rho(v) = k$ (see Fig. 10.3). We shall call a pair (A, B) of labelled trees a **linkage** if B can be obtained from A by the above construction. Our aim is to count the total number of possible linkages (A, B).

Since A may be chosen in any of $T(n, k-1)$ ways, and B is uniquely defined by the edge $\{w, z\}$ which may be chosen in $(n-1)-(k-1) = n-k$ ways, the total number of linkages (A, B) is clearly $(n-k)T(n, k-1)$. On the other hand, let B be a labelled tree in which

$\rho(v) = k$, and let $T_1, ..., T_k$ be the subtrees obtained from B by removing the vertex v and every edge incident to v; then we can obtain a labelled tree A for which $\rho(v) = k-1$ by removing from B just one of these edges ($\{v, w_i\}$, say, where w_i lies in T_i) and joining w_i to any vertex u of any other subtree T_j (see Fig. 10.4). It is clear

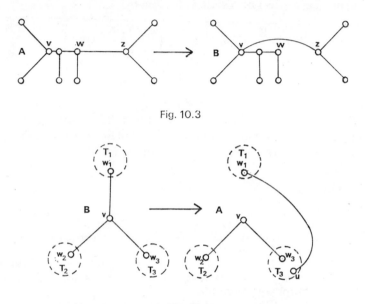

Fig. 10.3

Fig. 10.4

that the corresponding pair (A, B) of labelled trees is a linkage, and that all linkages may be obtained in this way. Since B may be chosen in $T(n, k)$ ways, and the number of edges joining w_i to vertices in any other T_j is $(n-1)-n_i$ (where n_i denotes the number of vertices of T_i), it follows that the total number of linkages (A, B) is

$$T(n, k)\{(n-1-n_1)+...+(n-1-n_k)\} = (n-1)(k-1)T(n, k),$$

since $n_1+...+n_k = n-1$.

We have thus shown that

$$(n-k)T(n, k-1) = (n-1)(k-1)T(n, k).$$

On iterating this result, and using the obvious fact that $T(n, n-1) = 1$, we deduce immediately that

$$T(n, k) = \binom{n-2}{k-1}(n-1)^{n-k-1}.$$

On summing over all possible values of k, it follows that the number $T(n)$ of labelled trees on n vertices is given by

$$T(n) = \sum_{k=1}^{n-1} T(n, k) = \sum_{k=1}^{n-1} \binom{n-2}{k-1}(n-1)^{n-k-1}$$

$$= \{(n-1)+1\}^{n-2} = n^{n-2}.//$$

COROLLARY 10B. *The number of spanning trees of K_n is n^{n-2}.*

Proof. To every labelled tree on n vertices there corresponds (in a unique way) a spanning tree of K_n; conversely, every spanning tree of K_n gives rise to a unique labelled tree on n vertices.//

Exercises

(*10a*) Show that there are exactly $2^{\frac{1}{2}n(n-1)}$ labelled simple graphs on n vertices; how many of these have exactly m edges?

(*10b*) Verify directly that there are exactly 125 labelled trees on five vertices.

(*10c*) Show that a given simple graph G on n vertices can be labelled in exactly $n!/g$ different ways, where g denotes the order of the automorphism group of G.

(*10d*) Show that if $\rho_1, \ldots, \rho_n$ are given positive integers, then there exists a labelled tree on n vertices in which the degree of v_k is ρ_k (for each k) if and only if $\Sigma \rho_k = 2(n-1)$.

(*10e*) Show that the number of edge-labelled trees on n vertices (in which the edges are labelled, rather than the vertices) is n^{n-3} ($n \geq 3$).

(*10f*) Show that if n is large, the probability that a given vertex of a tree with n vertices is a terminal vertex is approximately e^{-1}.

(**10g*) If $T(n)$ denotes the number of labelled trees on n vertices, show (by counting the number of ways of joining together a labelled tree on k vertices and one on $n-k$ vertices) that

$$2(n-1)T(n) = \sum_{k=1}^{n-1} \binom{n}{k}k(n-k)T(k)T(n-k);$$

deduce the identity

$$\sum_{k=1}^{n-1} \binom{n}{k}k^{k-1}(n-k)^{n-k-1} = 2(n-1)n^{n-2}.$$

(*$10h$) (**Matrix-tree theorem**) It can be shown that the number of spanning trees of a labelled simple graph G with vertex set $\{v_1, \ldots, v_n\}$ is equal to the cofactor of any entry of the $n \times n$ matrix $M = (m_{ij})$, in which $m_{ii} = \rho(v_i)$, $m_{ij} = -1$ if v_i and v_j are adjacent, and $m_{ij} = 0$ otherwise; verify this result when G is a complete graph, and show how you would extend it to graphs containing loops or multiple edges.

§11. SOME APPLICATIONS OF GRAPH THEORY

Although we are primarily concerned in this book with the *theory* of graphs, it is high time that we mentioned some possible applications. After all, most of the important advances in the subject arose as a result of attempts to solve particular 'practical' problems—Euler and the bridges of Königsberg, Cayley and the enumeration of saturated hydrocarbons, and various investigations into the colouring of maps, to name but three. Much of the present-day interest in the subject is due to the fact that, quite apart from being an elegant mathematical discipline in its own right, graph theory is playing an ever-increasing rôle in such a wide range of subjects as electrical engineering and linguistics, operational research and crystallography, probability and genetics, and sociology, geography and numerical analysis.

It is inappropriate in a book of this size to try to discuss a large number of applications in any kind of detail; for this, we refer the reader to the excellent account in Chapter 6 of Busacker and Saaty.[2] We wish simply to remind the reader of some of the applications we have already met, and to describe briefly what lies in store; this will be followed by a detailed discussion of two particular problems.

We have already seen how graphs and digraphs can be used to represent many situations; these include *marketing* (exercise *3f*) using a bipartite graph to depict various factories and the markets to which their goods are sent; *games* (exercise *1a(v)*) with the vertices representing various stages of the game and the arcs representing possible moves; *social relationships* (the handshaking lemma and exercises *1a(iii)*, *3f*, *3j* and *7i*); *electrical networks* and *road maps* (§1); *chemical compounds* (§10, and exercises *1b* and *9c*); and *puzzles* (the second paragraph of §6, and exercises *6d, 6g, 7d*, and *7i*).

Later on, we shall be mentioning such applications as the use of planar graphs in the study of printed circuits (§13), chromatic polynomials in timetabling problems (§21), and transversal theory in the construction of latin squares and in group theory (§27). We shall also be discussing in some detail the use of digraphs in the study of Markov chains (§24) and in the problem of finding maximal flows in transportation networks (§29).

The remainder of this section is devoted to a more detailed study of two particular applications, the first involving the minimal connector problem, the second being rather more trivial.

(*i*) Let us suppose that we wish to build a railway network connecting *n* given cities in such a way that a passenger can travel from any city to any other. If we assume that for economic reasons the amount of track used must be a minimum, then it is clear that the graph formed by taking the *n* cities as vertices and the connecting

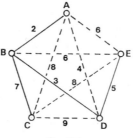

Fig. 11.1

rails as edges must be a tree. The problem is to find an algorithm for deciding which of the n^{n-2} possible trees connecting these cities uses the least amount of track, assuming that the distances between pairs of cities are known (see Fig. 11.1).

We can state a slightly more general form of this problem in graph-theoretic terms as follows: let G be a connected graph, and suppose that to each edge e of G is assigned a non-negative real number $\mu(e)$, called its **measure;** then we wish to find an algorithm for obtaining a spanning tree T whose **measure-sum** $M(T) = \sum \mu(e)$ is as small as possible, the sum being taken over the edges of T. This problem is known as the **minimal connector problem,** and reduces to the previous problem by taking G to be the graph K_n, and the measure of an edge to be the distance between the corresponding pair of cities; the following theorem gives us the required algorithm, known generally as **Kruskal's algorithm.**

THEOREM 11A. *Let G be a connected graph with n vertices; then the following construction gives a solution of the minimal connector problem*: (*i*) *let e_1 be an edge of G of smallest measure*; (2) *define $e_2, e_3, ..., e_{n-1}$ inductively by choosing at each stage an edge (not previously chosen) of smallest measure, subject to the condition that it forms no circuit with the previous edges e_i. The required spanning tree is then the subgraph T of G whose edges are $e_1, ..., e_{n-1}$.*

Remark. The reader should verify that if G is the graph shown in Fig. 11.1, then this construction yields: $e_1 = AB$, $e_2 = BD$, $e_3 = DE$, $e_4 = BC$.

Proof. The fact that T is a spanning tree of G follows immediately from statement (*ii*) of theorem 9A; it remains only to show that the measure-sum of T is a minimum. In order to do this, we suppose that S is a spanning tree of G with the property that $M(S) < M(T)$. If e_k is the first edge in the above sequence which does not lie in S, then the subgraph of G formed by adding e_k to S contains a unique circuit C containing the edge e_k. Since C clearly contains an edge e lying in S but not in T, it follows that the subgraph obtained from S on replacing e by e_k is still a spanning tree (S', say). But by the construction, $\mu(e_k) \leqq \mu(e)$, and so $M(S') \leqq M(S)$, and S' has one more edge in common with T than S; it follows on repeating this procedure that we can change S into T, one step at a time, with the measure-sum decreasing at each stage; hence $M(T) \leqq M(S)$, giving us the required contradiction.//

A problem which superficially resembles the minimal connector problem is the well-known **travelling salesman problem** which asks for an algorithm for the solution of the following problem: a man wishes to visit n given cities; how can he visit them all at least once in such a way that the total distance travelled is as small as possible? (In Figure 11.1, the shortest possible route is $A \rightarrow B \rightarrow D \rightarrow E \rightarrow C \rightarrow A$, giving a total distance of 26, as can be seen by inspection.) The practical applications of this problem are very wide-ranging, but unfortunately no general algorithm for its solution is known at present.

(*ii*) A puzzle which has been popular recently and which has been marketed under the name of 'Instant Insanity', concerns four cubes whose faces are coloured red, blue, green and yellow in such a way that each cube contains at least one face of each colour (as in Fig. 11.2); the problem is to pile these cubes up on top of each other in such a way that each of the four 4×1 sides of the resulting square prism shows a face of each colour.

In order to solve this problem, we represent each cube by a graph on four vertices, one vertex corresponding to each colour; in each such graph, two vertices are adjacent if and only if the cube in question has the corresponding colours on opposite faces. The graphs corresponding to the cubes of Fig. 11.2 are shown in Fig. 11.3.

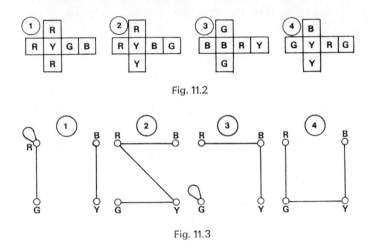

Fig. 11.2

Fig. 11.3

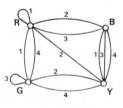

Fig. 11.4

We shall find it convenient to superimpose these graphs to form a new graph G (Fig. 11.4). Since every solution of the puzzle has two faces of each colour on each of the two pairs of opposite sides of the 4×1 prism, it is not difficult to see that the required solution is obtained by finding two edge-disjoint subgraphs H_1 and H_2 of G which are regular of degree two and which contain exactly one edge of each number (the subgraphs corresponding to our particular example are shown in Fig. 11.5). H_1 and H_2 then represent the colours appearing

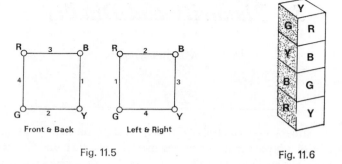

Fig. 11.5   Fig. 11.6

on the front-and-back and on the left-and-right sides of the 4×1 prism; the solution can now be read off from these subgraphs (Fig. 11.6).

Exercises

(*11a*) Use the construction of theorem 11A to find a spanning tree, with smallest possible measure-sum, of the graph shown in Fig. 11.7.

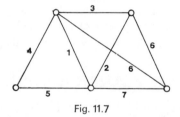

Fig. 11.7

(*11b*) Find an alternative algorithm for the minimal connector problem, involving the removal of edges of greatest measure from the graph G. Show also that if every edge has the same measure, then this algorithm gives a method for constructing a spanning tree of G.

(*11c*) Use theorem 11A to show that if V is a finite-dimensional vector space, then any two bases of V contain the same number of elements.

(*11d*) Suppose that in the travelling salesman problem all of the n cities are contained in a square of side k; by splitting up this square into m parallel narrow strips, show that the total distance travelled need not exceed $k(m+3+[n/m])$. By suitably choosing m, show that for large n, this distance is not greater than approximately $2k\sqrt{n}$.

(*11e*) Show that in the problem of the coloured cubes, there are 41 472 different ways of forming a 4×1 square prism; show, however, that in our particular example, only one of these ways leads to a solution of the problem.

5 Planarity and Duality

Flattery will get you nowhere.
Popular saying

We now embark upon a study of topological graph theory, in which the study of graph theory becomes inextricably tied up with topological notions such as planarity, genus, etc. In §4 it was proved that every graph can be embedded (i.e. drawn without crossings) in three-dimensional space; we now investigate conditions under which a graph can be embedded in the plane and other surfaces. In §12 we prove the existence of graphs which are not planar, and state Kuratowski's famous characterization of planar graphs. Euler's theorem relating the numbers of vertices, edges and faces of a plane graph is then proved in §13, and generalized to graphs embedded in other surfaces in §14. The remainder of the chapter is devoted to a study of duality from three points of view—the geometric approach, the circuit-cutset approach and the Whitney-dual approach.

§12. PLANAR GRAPHS

A **plane graph** is a graph drawn in the plane in such a way that no two edges (or rather, the curves representing them) intersect geometrically except at a vertex to which they are both incident; a **planar graph** is one which is isomorphic to a plane graph. In the language of §4, this amounts to saying that a graph is planar if it can be embedded in the plane, and that any such embedding is a plane graph; for example, all three graphs in Fig. 12.1 are planar, but only the second and third are plane.

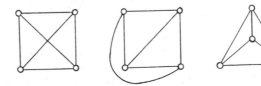

Fig. 12.1

One question which arises from the example just given and from exercise *4c* is whether a planar graph can always be drawn in the plane in such a way that all of its edges are represented by straight lines. Although this is clearly false for graphs containing loops or multiple edges, it is in fact true for simple graphs, as was proved by Fáry in 1948. The interested reader should consult Busacker and Saaty[2] for further details.

Not all graphs are planar, as the following theorem shows:

THEOREM 12A. *K_5 and $K_{3,3}$ are non-planar.*

Remark. We shall be giving two proofs of this result. The first one, which is presented here, depends on the Jordan curve theorem in the form in which it was given in §4; the second proof, which we defer until the next section, will appear as a corollary of Euler's theorem.

Proof. Suppose that K_5 is planar. Since K_5 contains a circuit of length five (which we shall take as $v \to w \to x \to y \to z \to v$), any plane embedding can without loss of generality be assumed to contain this circuit drawn in the form of a regular pentagon (as in Fig. 12.2). By the Jordan curve theorem, the edge $\{z, w\}$ must lie

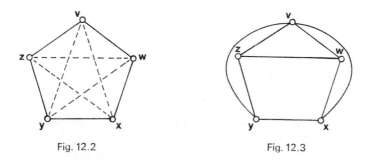

Fig. 12.2   Fig. 12.3

either wholly inside the pentagon or wholly outside it (the third possibility, namely that the edge has a point in common with the pentagon, does not arise since we are assuming a plane embedding); we shall deal with the case in which $\{z, w\}$ lies inside the pentagon— the other case is similar and will be left to the reader. Since the edges $\{v, x\}$ and $\{v, y\}$ do not cross the edge $\{z, w\}$, they must both lie outside the pentagon; the situation is now as in Fig. 12.3. But the edge $\{x, z\}$ cannot cross the edge $\{v, y\}$ and so must lie inside the pentagon, and similarly the edge $\{w, y\}$ must also lie inside the pentagon. Since the edges $\{w, y\}$ and $\{x, z\}$ must then cross, we obtain the required contradiction.

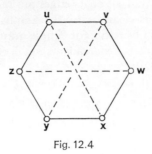

Fig. 12.4

A similar, but easier, argument is used to show that $K_{3,3}$ is non-planar; we simply draw a hexagonal circuit $u \to v \to w \to x \to y \to z \to u$ as in Fig. 12.4, and show (using the Jordan curve theorem) that two of the edges $\{u, x\}$, $\{v, y\}$, $\{w, z\}$ must both lie inside or outside the hexagon, and hence must cross.//

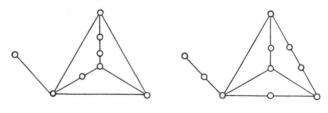

Fig. 12.5

It is clear that every subgraph of a planar graph is planar, and that every graph which contains a non-planar graph as a subgraph must itself be non-planar; from this we immediately deduce that any graph which contains K_5 or $K_{3,3}$ as a subgraph cannot be planar. It turns out that K_5 and $K_{3,3}$ are essentially the only non-planar graphs, in the sense that every non-planar graph 'contains' one of them. To make this statement more precise, we need the concept of 'homeomorphic graphs'.

Two graphs are **homeomorphic** (or **identical to within vertices of degree two**) if they can both be obtained from the same graph by inserting new vertices of degree two into its edges; for example, the graphs shown in Fig. 12.5 are homeomorphic, and so are any two circuit graphs. Note that the homeomorphism of graphs is an equivalence relation.

It is clear that the introduction of the term 'homeomorphic' is merely a technicality—the insertion or deletion of vertices of degree two is irrelevant to any considerations of planarity. However, it enables us to state the following important result which is known as **Kuratowski's theorem** and which gives a necessary and sufficient condition for a graph to be planar.

THEOREM 12B (Kuratowski 1930). *A graph is planar if and only if it contains no subgraph homeomorphic to K_5 or $K_{3,3}.//*

The proof of Kuratowski's theorem is rather long and involved, and for this reason we have decided to omit it (see references 1, 2, 3 or 6 in the bibliography). We shall however use Kuratowski's theorem to obtain another criterion for planarity.

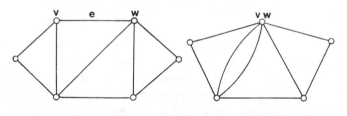

Fig. 12.6

We shall need a couple of preliminary definitions: an **edge-contraction** in a graph is obtained by taking an edge e (with incident vertices v and w, say) and 'contracting' it—in other words, removing e and identifying v and w in such a way that the resulting vertex is incident to those edges (other than e) which were originally incident to v or w (see Fig. 12.6). A **contraction** of a graph G is then a graph which results from G after a succession of edge-contractions. Note that K_5 is a contraction of the Petersen graph (by contracting the five edges which connect the inner circuit to the outer one); we can also express this by saying that the Petersen graph is **contractible** to K_5.

THEOREM 12C. *A graph is planar if and only if it contains no subgraph which is contractible to K_5 or $K_{3,3}$.*

★ *Proof.* ⇐ Assume first that the graph G is non-planar; then by Kuratowski's theorem, G contains a subgraph H which is homeomorphic to K_5 or $K_{3,3}$. On contracting those edges of H which are incident to vertices of degree two, we see immediately that H is contractible to K_5 or $K_{3,3}$.

⇒ Now assume that G contains a subgraph H which is contractible to $K_{3,3}$, and let the vertex v of $K_{3,3}$ arise from the contraction of the subgraph H_v of H (see Fig. 12.7). The vertex v is incident in $K_{3,3}$

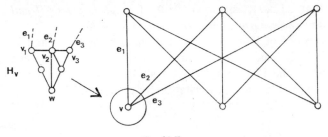

Fig. 12.7

to three edges e_1, e_2 and e_3; when regarded as edges of H, these edges are incident to three (not necessarily distinct) vertices v_1, v_2 and v_3 of H_v. If v_1, v_2 and v_3 are distinct, we can find a vertex w of H_v and three chains from w to these vertices, these chains intersecting only at w. (A similar construction can be made if the vertices are not distinct, the chains degenerating in this case to single vertices). It follows that we can replace the subgraph H_v by a vertex w and three chains leading out of it. If this construction is carried out for each vertex of $K_{3,3}$, and the resulting chains joined up with the corresponding edges of $K_{3,3}$, the resulting subgraph will clearly be homeomorphic to $K_{3,3}$, showing (by Kuratowski's theorem) that G is non-planar (see Fig. 12.8).

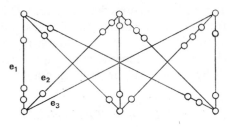

Fig. 12.8

A similar argument can be carried out if G contains a subgraph which is contractible to K_5; we shall leave the details as an exercise for the reader (see exercise *12l*).//★

Exercises

(*12a*) Three neighbours use the same water, oil and treacle wells (see Carroll[12]). Unfortunately they dislike each other so much that in order to avoid meeting they want to find non-crossing paths from each of their houses to each of the three wells. Can this be done?

(*12b*) For which values of n are the graphs G_n (as defined in exercise *2a*) planar?

(*12c*) Use Kuratowski's theorem to show that the Petersen graph is non-planar.

(*12d*) Let G be a planar graph with vertex-set $\{v_1, ..., v_n\}$, and let $p_1, ..., p_n$ be any n distinct points in the plane; give a heuristic argument to show that G can be embedded in the plane in such a way that the point p_i represents the vertex v_i for each i.

(*12e*) Prove the above statement that the homeomorphism of graphs is an equivalence relation. If two homeomorphic graphs have n_i vertices and m_i edges ($i = 1, 2$), show that $m_1 - n_1 = m_2 - n_2$.

(*12f*) Let G be a simple graph; show that if G is non-planar then so is its line-graph $L(G)$. If G is planar, is $L(G)$ necessarily planar?

(*12g*) Show that a planar graph can be embedded in the plane in only finitely many topologically distinct ways.

(*12h*) The **crossing-number** $cr(G)$ of a graph G is the smallest possible number of crossings occurring when the graph is drawn in the plane (where the word 'crossing' refers, as in §4, to the intersection of exactly two edges). Show that (*i*) the crossing-number of a planar graph is zero; (*ii*) $cr(K_5) = cr(K_{3,3}) = 1$. What is the crossing-number of the Petersen graph?

(**12i*) Let G be the graph formed by the vertices, edges, and principal diagonals of a $2n$-sided polygon ($n \geq 3$); show that $cr(G) = 1$. Can you find a corresponding result for a $(2n+1)$-sided polygon?

(**12j*) Show that if m and n are both even, then $cr(K_{m,n}) \leq \frac{1}{16}mn(m-2)(n-2)$, and obtain similar results if m and n are not necessarily even. (Hint: place the m vertices at equal intervals along the x-axis with an equal number on each side of the origin; place the other n vertices along the y-axis in a similar way—now count the crossings.)

(*12k*) For each of the following pairs of graphs, state whether the second can be obtained as a contraction of the first: (*i*) the Petersen graph, K_5; (*ii*) K_n, K_{n-1}; (*iii*) W_6, K_4; (*iv*) $K_{3,3}$, K_4; (*v*) the cube graph, W_5; (*vi*) the Petersen graph, W_6. Find a non-planar graph which is not contractible to K_5 or $K_{3,3}$; why does this not contradict theorem 12c?

(**12l*) Complete the proof of theorem 12c. (Warning: one part of the proof will need rather careful treatment.)

§13. EULER'S THEOREM ON PLANE GRAPHS

In this section we shall prove a theorem relating the numbers of vertices, edges and faces of a given connected plane graph G. Before defining exactly what is meant by a 'face' of G, we remind the reader that a point x of the plane is said to be 'disjoint from G' if x represents neither a vertex of G nor a point which lies on an edge of G.

If x is a point of the plane disjoint from G, we define the **face** (of G) **containing x** to be the set of all points of the plane which can be reached from x by a Jordan curve all of whose points are disjoint from G. Alternatively, we can say that two points x and y of the plane are equivalent if they are both disjoint from G and can be joined by a Jordan curve all of whose points are disjoint from G (Fig. 13.1);

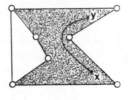

Fig. 13.1

this is an equivalence relation on the points of the plane disjoint from G, and the corresponding equivalence classes are called the **faces** of G. Note that one face is unbounded; it is called the **infinite face.** For example, if G is the graph of Fig. 13.2, then G has four

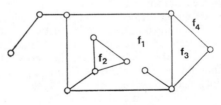

Fig. 13.2

faces, f_4 being the infinite face. Any reader who feels that our definition of a face is too pedantic, may safely rely on his intuition.

It is important to realize that there is nothing special about the infinite face—in fact, any face can be chosen as the infinite face. To see this, we use theorem 4B to map the graph onto the surface of a sphere; we now rotate the sphere so that the point of projection (i.e. the north pole) lies inside the face we want as the infinite face, and then project the graph down onto the plane which is tangent to the sphere at the south pole. The chosen face is now the infinite face. Fig. 13.3 shows a representation of the graph of Fig. 13.1 in which the infinite face is f_3. From now on, we shall feel free to talk interchangeably about graphs embedded in the plane and graphs drawn on the surface of a sphere.

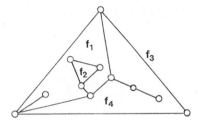

Fig. 13.3

We now state and prove **Euler's theorem**, which tells us that whatever plane embedding of a graph we take, the number of faces always remains the same and is given by a simple formula; an alternative proof will be outlined in exercise *13l*.

THEOREM 13A (Euler 17~~36~~50). *Let G be a connected plane graph, and let n, m and f denote respectively the number of vertices, edges and faces of G. Then*

$$n+f = m+2.$$

Proof. The proof is by induction on the number of edges of G. If $m = 0$, then $n = 1$ (since G is connected) and $f = 1$ (the infinite face); the theorem is therefore true in this case.

Now suppose that the theorem is true whenever G has $m-1$ edges and let us add to G a new edge e. Then either (*i*) e is a loop, in which case a new face is created, but the number of vertices remains unchanged, or (*ii*) e joins two distinct vertices of G, in which case one of the faces of G is split into two, increasing the number of faces by one, but leaving the number of vertices unchanged, or (*iii*) e is incident to only one vertex of G, in which case another vertex must be added, increasing the number of vertices by one, but leaving the number of faces unchanged. In each case the theorem remains true. Since these are the only possible cases, the theorem is proved.//

This result is often called 'Euler's polyhedron formula' since it relates the numbers of vertices, edges and faces of a convex polyhedron; this can easily be seen by projecting the polyhedron out onto the surface of its circumsphere, and then using theorem 4B. The resulting plane graph is a connected graph in which every face is bounded by a polygon—such a graph is called a **polyhedral graph** (see Fig. 13.4). For convenience we restate theorem 13A for such graphs.

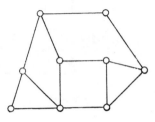

Fig. 13.4

COROLLARY 13B. *Let G be a polyhedral graph*; *then with the above notation*,

$$n+f = m+2.//$$

Euler's theorem can easily be extended to disconnected graphs:

COROLLARY 13C. *Let G be a plane graph with n vertices, m edges, f faces and k components*; *then*

$$n+f = m+k+1.$$

Proof. The result follows immediately on applying Euler's theorem to each component separately, remembering not to count the infinite face more than once.//

All of the results mentioned so far in this section apply to arbitrary plane graphs; we must now restrict ourselves to simple graphs.

COROLLARY 13D. *If G is a connected simple planar graph with n (≧3) vertices and m edges, then*

$$m \leqq 3n-6.$$

∠ M ≧ 3 ?

NB .

h = V

Proof. We can assume without loss of generality that G is a plane graph. Since every face is bounded by at least three edges, it follows on counting up the edges around each face that $3f \leq 2m$ (the factor 2 arising from the fact that every edge bounds at most two faces). We obtain the required result by combining this inequality with Euler's theorem.//

This corollary can be used to give an alternative proof of theorem 12A.

COROLLARY 13E. *K_5 and $K_{3,3}$ are non-planar.*

Proof. If K_5 is planar, then applying corollary 13D, we obtain $10 \leq 9$, which is clearly a contradiction. To show that $K_{3,3}$ is non-planar, we note that every face is bounded by at least four edges (see exercise 5g), and hence that $4f \leq 2m$ (i.e. $2f \leq 9$). But this is a contradiction, since Euler's theorem tells us that $f = 5$.//

A similar argument is used to prove the following theorem which will be useful when we come to study the colouring of graphs.

THEOREM 13F. *Every simple planar graph contains a vertex whose degree is at most five.*

Proof. Without loss of generality we can assume the graph to be plane and connected, and to contain at least three vertices. If every vertex has degree at least six, then with the above notation we have $6n \leq 2m$ (i.e. $3n \leq m$). It then follows immediately from corollary 13D that $3n \leq 3n - 6$, an obvious contradiction.//

We conclude this section with a few remarks on the 'thickness' of a graph. In electrical engineering, parts of networks are sometimes printed on one side of a non-conducting plate (and are called 'printed circuits'). Since the wires are not insulated, they cannot cross and the corresponding graphs must be planar. For a general network, it is of importance to know how many printed circuits are needed to complete the entire network; to this end, we define the **thickness** of a graph G (denoted by $t(G)$) to be the smallest number of planar graphs which can be superimposed to form G. Like the crossing-number, the thickness is a measure of how 'un-planar' a graph is; for example, the thickness of a planar graph is one, and of K_5 and $K_{3,3}$ is two.

As we shall see, a lower bound for the thickness of a graph may easily be obtained using Euler's theorem; what is rather surprising is that this rather trivial lower bound frequently turns out to be the correct value, as may be verified in special cases by direct construction. In deriving this lower bound, we shall use the symbols $[x]$ and $\{x\}$ to denote respectively the largest integer not greater than x and the smallest integer not less than x (so that, for example, $[3] = \{3\} = 3$; $[\pi] = 3$; $\{\pi\} = 4$); note that $\{x\} = -[-x]$.

THEOREM 13G. *Let G be a simple graph with n* (≥ 3) *vertices and m edges; then the thickness* $t(G)$ *of G satisfies the following inequalities*:

$$t(G) \geq \left\{\frac{m}{3n-6}\right\}; \quad t(G) \geq \left[\frac{m+3n-7}{3n-6}\right].$$

Proof. The first part is an immediate application of corollary 13D, the curly brackets arising from the fact that the thickness must be an integer. The second part follows from the first by using the easily-proved relation $\{a/b\} = [(a+b-1)/b]$ (a and b denoting positive integers). $/\!/$

Exercises

(*13a*)   Verify Euler's theorem for (*i*) W_n; (*ii*) the Platonic graphs; (*iii*) the graph formed by the vertices, edges and faces of an $n \times n$ chess-board; (*iv*) $K_{2,n}$.

(*13b*)   Redraw the graph of Fig. 13.1 with (*i*) f_1 as the infinite face; (*ii*) f_2 as the infinite face.

(*13c*)   Show that if G is a connected planar graph with girth r, then with the above notation, $(r-2)m \leq r(n-2)$; deduce that the Petersen graph is non-planar. Show moreover that if every face is bounded by a polygon with r sides, then the above inequality becomes an equality.

(*13d*)   Let G be a polyhedron (or polyhedral graph), all of whose faces are bounded by pentagons and hexagons; what can you say about the number of pentagonal faces? Show that if there are exactly three faces meeting at each vertex, then there must be exactly twelve pentagonal faces.

(*13e*)   Let G be a plane graph with less than twelve faces; show that if every vertex of G has degree at least three, then there is a face of G which is bounded by at most four edges.

(*13f*)   Let G be a connected cubic simple plane graph, and ϕ_n be the number of faces of G which are bounded by n edges. Show (by counting the number of edges and vertices of G) that

$$12 = 3\phi_3 + 2\phi_4 + \phi_5 - \phi_7 - 2\phi_8 - 3\phi_9 - \dots .$$

Deduce the last part of exercise *13d* as a special case, and prove also that G contains at least one face which is bounded by not more than five edges.

(13g) Let G be a simple graph with at least eleven vertices and let $\bar{G}$ denote its complement; show that G and $\bar{G}$ cannot both be planar. (In fact, a similar result can be proved with 'eleven' replaced by 'nine'.) Give an example of a graph G on eight vertices with the property that both G and $\bar{G}$ are planar.

(13h) Find the thickness of the Petersen graph.

(13i) Show that every non-planar graph is homeomorphic to a graph of thickness two.

(13j) Show that the thickness of K_n satisfies $t(K_n) \geq [\frac{1}{6}(n+7)]$. Show (by direct construction, and by using the result of exercise 13g) that equality holds if $n \leq 8$, but not if $n = 9$ or 10. (In fact, apart from these two exceptions, equality is known to hold for all $n \leq 51$ and for all n which are not of the form $6k+4$, and is thought to hold for these values as well.)

(*13k) Obtain a lower bound for the thickness of a graph with m edges, n vertices and with girth r. Show that if m is even ($=2k$), then $t(K_{m,n}) \leq k$, and deduce that the lower bound obtained for $t(K_{m,n})$ is an equality if $n > \frac{1}{2}(m-2)^2$. Obtain a corresponding result if m is odd.

(*13l) Let G be a polyhedral graph, and let W denote the circuit subspace of G; show that the polygons which bound the finite faces of G form a basis for W, and deduce corollary 13B. Show how this result can be extended to arbitrary plane graphs.

§14. GRAPHS ON OTHER SURFACES

★ In the previous two sections we considered graphs drawn in the plane or (equivalently) on the surface of a sphere. We shall now make a few remarks on the embedding of graphs on other surfaces—for example, the torus. It is easy to see that K_5 and $K_{3,3}$ can be drawn without crossings on the surface of a torus, and it is natural to ask whether there are analogues of Euler's theorem and Kuratowski's theorem for graphs drawn on such surfaces.

The torus can be thought of as a sphere with one 'handle' (Fig. 14.1); more generally, a surface is said to be of genus g if it is topologically homeomorphic to a sphere with g handles. (The reader to whom these terms are unfamiliar can think of graphs drawn on the surface of a doughnut with g holes in it.) Thus the genus of a sphere is zero, and of a torus is one.

A graph which can be drawn without crossings on a surface of genus g, but not on one of genus $g-1$, is called a **graph of genus g**. Thus K_5 and $K_{3,3}$ are graphs of genus one (also called **toroidal graphs**). We must check that the genus of a graph is well-defined.

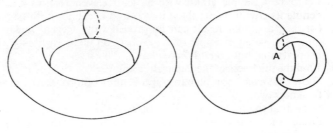

Fig. 14.1

THEOREM 14A. *The genus of a graph is well-defined, and does not exceed the crossing-number.*

Proof. To show that the genus is well-defined, it is sufficient to find an upper bound for it. This is done by drawing the graph on the surface of a sphere in such a way that the number of crossings is as small as possible (and is therefore equal to the crossing-number c). At each crossing, we construct a 'bridge' (see Fig. 1.2, p. 1) and run one edge over the bridge and the other under it. Since each bridge can be regarded as a handle, we have embedded the graph on the surface of a sphere with c handles; it follows that the genus is well-defined, and does not exceed c.//

At the time of writing there is no known analogue of Kuratowski's theorem for surfaces of genus g. It has not even been conclusively shown whether or not there exists, for each value of g, a finite collection of 'forbidden' subgraphs of genus g, corresponding to the forbidden subgraphs K_5 and $K_{3,3}$ for graphs of genus zero. In the case of Euler's theorem, however, we are more fortunate since there is a natural generalization for graphs of genus g. In this generalization, a face of a graph of genus g is defined in the obvious way, namely in terms of Jordan curves drawn on the surface.

THEOREM 14B. *Let G be a connected graph of genus g, with n vertices, m edges and f faces. Then $n+f = m+(2-2g)$.*

Sketch of proof. We shall outline the main steps in the proof, leaving the details to the reader (exercise *14b*).

(*i*) Without loss of generality, we may assume that G is drawn on the surface of a sphere with g handles. We can also assume that the curves A (see Fig. 14.1) at which the handles meet the sphere are in fact circuits of G (by shrinking those circuits which contain these curves in their interior).

(*ii*) We next disconnect each handle at one end, in such a way that the handle has a free end E and the sphere has a corresponding hole H. We may assume that the circuit corresponding to the end of the handle appears at both the free end E and the hole H, since the additional vertices and edges required for this exactly balance each other, leaving $n-m+f$ unchanged.

(*iii*) We complete the proof by telescoping each of these handles, leaving a sphere with $2g$ holes in it. Note that this telescoping process does not change the value of $n-m+f$. But for a sphere, $n-m+f=2$, and hence for a sphere with $2g$ holes in it, $n-m+f=2-2g$. The result now follows immediately.//

COROLLARY 14C. *The genus $g(G)$ of a simple graph G with n (≥ 4) vertices and m edges satisfies the inequality*

$$g(G) \geq \{\tfrac{1}{6}(m-3n)+1\}.$$

Proof. Since every face is bounded by at least three edges, we have (as in the proof of corollary 13D) $3f \leq 2m$. The result follows by substituting this inequality into theorem 14B, and using the fact that the genus of a graph must be an integer.//

As in the case of the thickness of a graph, little is known about the problem of finding the genus of an arbitrary graph. The usual method is to use corollary 14C to obtain a lower bound for the genus, and then to try to obtain the required embedding by direct construction.

One case of particular historical importance is that of the genus of the complete graphs. Corollary 14C tells us that the genus of K_n satisfies

$$g(K_n) \geq \{\tfrac{1}{6}(\tfrac{1}{2}n(n-1)-3n)+1\} = \{\tfrac{1}{12}(n-3)(n-4)\}.$$

Heawood conjectured in 1890 that the inequality just obtained is in fact an equality, and this was finally proved in 1968 by Ringel and Youngs after a long and difficult struggle.

THEOREM 14D (Ringel and Youngs 1968). $g(K_n) = \{\tfrac{1}{12}(n-3)(n-4)\}$.

Remark. This will not be proved here; the reader should consult Harris[8] for a discussion of this theorem.//

Further results concerning the embedding of graphs on other surfaces can be found in Harary.[1]

Exercises

(*14a*) Show that the surface of a torus can be regarded as a rectangle in which opposite edges have been identified (see Fig. 14.2). Find a similar representation for a sphere with g handles.

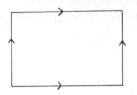

Fig. 14. 2

(*14b*) Complete the details of the proof of theorem 14B.

(*14c*) Can K_5 and $K_{3,3}$ be embedded on the surface of a Möbius strip? Find an analogue of Euler's theorem for this surface.

(*14d*) Give an example of a graph of genus two.

(*14e*) Show that there is no complete graph of genus seven; what is the next integer which is not the genus of any complete graph?

(*14f*) Obtain a lower bound for the genus of a graph with girth r; deduce that the genus of $K_{m,n}$ satisfies

$$g(K_{m,n}) \geqq \{\tfrac{1}{4}(m-2)(n-2)\}.$$

(In fact, Ringel has shown that this is an equality.)

(**14g*) The **toroidal thickness** $t_1(G)$ of a graph G is the smallest number of toroidal graphs which can be superimposed to form G; show that

$$t_1(K_n) \geqq [\tfrac{1}{6}(n+4)].$$

How would you define more complicated types of thickness, and what form would the corresponding results for K_n take? ★

§15. DUAL GRAPHS

In theorems 12B and 12C we gave necessary and sufficient conditions for a graph to be planar, namely that it contains no subgraph which is homeomorphic or contractible to K_5 or $K_{3,3}$. Our aim is now to discuss conditions of a rather different kind; these will involve the concept of duality.

Given a plane graph G, we shall construct another graph G^* called the (**geometric**)-**dual** of G. The construction is in two stages: (*i*) inside each face F_i of G we choose a point v_i^*—these points are the vertices of G^*; (*ii*) corresponding to each edge e of G we draw a line e^* which crosses e (but no other edge of G), and joins the vertices $v_i{}^*$ which lie in the two (not necessarily distinct) faces F_i adjoining e—these lines are the edges of G^*. This procedure is illustrated in Fig. 15.1, the vertices $v_i{}^*$ being represented by crosses, the edges e of G by solid lines and the edges e^* of G^* by dashed lines. Note that a terminal vertex of G gives rise to a loop of G^*, as does any isthmus; note also that if two faces of G have more than one edge in common, then G^* contains multiple edges.

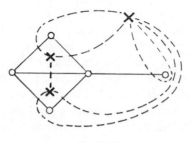

Fig. 15.1

It is clear that any two graphs formed from G in this way must be isomorphic; this is why we called G^* 'the dual of G' instead of 'a dual of G'. On the other hand, it should be pointed out that if G is isomorphic to H, it does not necessarily follow that G^* is isomorphic to H^*; an example which demonstrates this is given in exercise *15c*.

If G is not only plane, but connected as well, then G^* is plane and connected and there are simple relations connecting the numbers of vertices, edges and faces of G and G^*.

LEMMA 15A. *Let G be a plane connected graph with n vertices, m edges and f faces, and let G^*, its geometric-dual, have n^* vertices, m^* edges and f^* faces; then $n^* = f$, $m^* = m$ and $f^* = n$.*

Proof. The first two relations are direct consequences of the definition of G^*; the third relation follows immediately on substituting these two relations into Euler's theorem applied to both G and G^*.//

Since the dual G^* of a plane graph G is also a plane graph, we can repeat the construction described above to form the dual of G^* (denoted by G^{**}); if G is connected, then the relationship between G^{**} and G is particularly simple, as we now show.

D

THEOREM 15B. *Let G be a plane connected graph; then G** is isomorphic to G.*

Proof. The result follows almost immediately from the fact that the construction which gives rise to G^* from G can be reversed to give G from G^*; for example, in Fig. 15.1 the graph G is the dual of the graph G^*. We need to check only that a face of G^* cannot contain more than one vertex of G—it certainly contains at least one—and this follows immediately from the relations $n^{**} = f^* = n$, where n^{**} denotes the number of vertices of G^{**}.//

If G now denotes any planar graph, then a dual of G can be defined by taking any plane embedding and forming its geometric-dual, but uniqueness does not in general hold. Since duals have been defined only for planar graphs, it is trivially true to say that a graph is planar if and only if it has a dual; on the other hand, if we are given an arbitrary graph we have no way of telling from the above whether or not it is planar. It is obviously desirable to find a definition of duality which generalizes the geometric-dual and at the same time enables us (in principle, at least) to determine whether or not a given graph is planar.

One such definition exploits the relationship under duality between the circuits and cutsets of a planar graph G. We shall first describe this relationship and then use it to obtain the definition we seek; an alternative definition will be given in the following section.

THEOREM 15C. *Let G be a planar graph and G* be a geometric-dual of G; then a set of edges in G forms a circuit in G if and only if the corresponding set of edges of G* forms a cutset in G*.*

Proof. We can assume without loss of generality that G is a connected plane graph. If C is a circuit in G, then C encloses one or more of the finite faces of G, and thus contains in its interior a non-empty set S of vertices of G^*. It follows immediately that those edges of G^* which cross the edges of C form a cutset of G^* whose removal disconnects G^* into two subgraphs, one with vertex-set S and the other containing those vertices which do not lie in S (see Fig. 15.2). The converse implication is proved by simply reversing this argument.//

COROLLARY 15D. *A set of edges of G forms a cutset in G if and only if the corresponding set of edges of G* forms a circuit in G*.*

Proof. The result follows immediately on applying theorem 15c to G^* and using theorem 15b.//

Using theorem 15c as motivation, we can now give an abstract definition of duality; note that this definition does not invoke any special properties of planar graphs, but concerns only the relationship between two graphs.

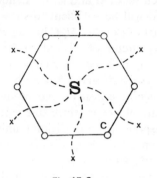

Fig. 15.2

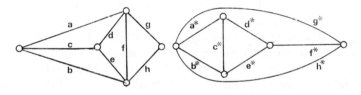

Fig. 15.3

We shall say that a graph G^* is an **abstract-dual** of a graph G if there is a one-one correspondence between the edges of G and those of G^* with the property that a set of edges of G forms a circuit in G if and only if the corresponding set of edges of G^* forms a cutset in G^*. For example, Fig. 15.3 shows a graph and its abstract-dual, with corresponding edges sharing the same letter.

It is clear from theorem 15c that the concept of an abstract-dual generalizes that of a geometric-dual, in the sense that if G is a planar graph and G^* is a geometric-dual of G then G^* is an abstract-dual of G; what we should like to be able to do is to obtain analogues for abstract-duals of some of the results on geometric-duals. We shall be content with just one of these here—the analogue for abstract-duals of theorem 15B.

THEOREM 15E. *If G* is an abstract-dual of G, then G is an abstract-dual of G*.*

Remark. Note that we do not require that G should be connected.

Proof. Let C be a cutset of G and let C^* denote the corresponding set of edges of G^*; it will be sufficient to show that C^* is a circuit of G^*. By the first part of exercise 5i, C has an even number of edges in common with any circuit of G, and so C^* must have an even number of edges in common with any cutset of G^*. It follows from the second part of exercise 5i that C^* must be either a single circuit in G^* or an edge-disjoint union of two or more circuits; but the second possibility cannot occur, since one can show similarly that circuits in G^* correspond to edge-disjoint unions of cutsets in G, and so C would then be an edge-disjoint union of two or more cutsets, rather than just a single cutset.//

Although the definition of an abstract-dual seems at first sight rather strange, it turns out to have the properties required of it. We saw in theorem 15C that a planar graph has an abstract-dual (e.g. any geometric-dual), and we now show that the converse result is true, namely that any graph which has an abstract-dual must be planar. In other words, we now have an abstract definition of duality which generalizes the geometric-dual and which characterizes planar graphs. It will turn out, in fact, that the definition of an abstract-dual is a natural consequence of the study of duality in matroid theory (see §**32**).

THEOREM 15F. *A graph is planar if and only if it has an abstract-dual.*

Remark. There are several proofs of this result. We shall be presenting a particularly simple one (due to T. D. Parsons) which uses Kuratowski's theorem.

★ *Sketch of proof.* As mentioned above, it is sufficient to prove that if G is a graph which has an abstract-dual G^*, then G is planar. The proof is in four steps:

(*i*) We note first that if an edge e is removed from G, then the abstract-dual of the remaining graph may be obtained from G^* by simply contracting the corresponding edge e^*. On repeating this procedure, it follows immediately that if G has an abstract-dual, then so does any subgraph of G.

(*ii*) We next observe that if G has an abstract-dual, and G' is homeomorphic to G, then G' also has an abstract-dual. This follows from the fact that the insertion or removal in G of a vertex of degree two results in the addition or deletion of a 'multiple edge' in G^*.

(*iii*) The third step is to show that neither K_5 nor $K_{3,3}$ has an abstract-dual. If G^* is a dual of $K_{3,3}$, then since $K_{3,3}$ contains only circuits of length four or six, and no cutsets consisting of only two edges, it follows that G^* contains no multiple edges, and that every vertex of G^* must have degree at least four. Hence G^* must contain at least five vertices, and thus at least $\frac{1}{2}.5.4 = 10$ edges, which is a contradiction. The argument for K_5 is similar, and will be left to the reader.

(*iv*) Suppose, now, that G is a non-planar graph which has an abstract-dual G^*. Then by Kuratowski's theorem, G contains a subgraph H homeomorphic to K_5 or $K_{3,3}$. It follows from (*i*) and (*ii*) that H, and hence also K_5 or $K_{3,3}$, must have an abstract-dual, contradicting (*iii*).//★

Exercises

(*15a*) Show that the dual of a wheel is a wheel; can you find any other graphs which are **self-dual** (i.e. isomorphic to their geometric-duals)?

(*15b*) Find the duals of the Platonic graphs.

(*15c*) Show that the graphs in Fig. 15.4 are isomorphic, but that their geometric-duals are non-isomorphic.

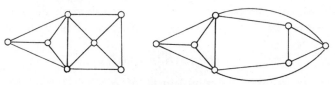

Fig. 15.4

(*15d*) Show that if a planar graph G is disconnected, then its geometric double-dual G^{**} is not isomorphic to G.

(*15e*) Under what conditions is the geometric-dual of a planar graph actually a simple graph?

(*15f*) Can you find a plane graph with five faces, with the property that any two faces have an edge in common?

(*15g*) Let G be a plane graph; use exercise *5g* and corollary *6c* to show that if G is bipartite then its geometric-dual G^* is Eulerian, and conversely.

(*15h) Let G be a connected plane graph which is regular of degree r and contains at least three vertices; if its dual G^* is also regular (of degree r^*, say) show that $(r-2)(r^*-2) < 4$. Show that the Platonic graphs satisfy this relation, and find every other graph which satisfies it.

(*15i) Show that if G^* is an abstract-dual of G, then any spanning forest of G corresponds to the complement of a spanning forest of G^*. Also find an expression for the cutset rank of a subgraph of G^* in terms of the cutset rank of the corresponding subgraph of G.

(*15j) A graph G^* is said to be an **algebraic-dual** of a graph G if there is a one-one correspondence between the edges of G and those of G^* with the property that elements of the circuit subspace of G correspond to elements of the cutset subspace of G^*, and conversely. Show that G^* is an algebraic-dual of G if and only if G^* is an abstract-dual of G.

§16. THE WHITNEY-DUAL

★ In the previous section we gave an abstract definition of duality which can be used to characterize planar graphs. We conclude this chapter with a short section describing an alternative definition of duality, namely that of a Whitney-dual. Although this definition will seem rather artificial at first sight, we are presenting it here for historical reasons since Whitney was the first graph theorist to characterize planar graphs by considering their duals.

We shall say that a graph G^* is a **Whitney-dual** of a graph G if there is a one-one correspondence between the edges of G and those of G^* with the property that if H is any subgraph of G with the same vertex-set as that of G, then the corresponding subgraph H^* of G^* satisfies

$$\gamma(H)+\kappa(\tilde{H}^*) = \kappa(G^*),$$

where $\tilde{H}^*$ denotes the complement of H^* in G^* (i.e. the graph obtained from G^* by removing the edges of H^*), and γ and κ are defined as in §9.

Some trivial consequences of this definition are given in the following theorem:

THEOREM 16A. *With the above notation,*

$$\gamma(G) = \kappa(G^*); \quad \gamma(G^*) = \kappa(G); \quad \kappa(H)+\gamma(\tilde{H}^*) = \kappa(G).$$

Proof. The first equation follows from the definition of the Whitney-dual by putting $H = G$, and observing that $\tilde{H}^*$ is a null graph and $\kappa(\tilde{H}^*) = 0$. The second equation follows from the first, and the equalities

$$\kappa(G^*) + \gamma(G^*) = m(G^*) = m(G) = \kappa(G) + \gamma(G),$$

where $m(G)$, $m(G^*)$ denote the number of edges of G and G^*, respectively.

For the third equation, we have

$$
\begin{aligned}
\kappa(H) + \gamma(\tilde{H}^*) &= m(H) - \gamma(H) + m(\tilde{H}^*) - \kappa(\tilde{H}^*) \\
&= m(H) + m(\tilde{H}) - \kappa(G^*) \quad (\text{since } m(\tilde{H}^*) = m(\tilde{H})) \\
&= m(G) - \gamma(G) = \kappa(G), \quad \text{as required.} //
\end{aligned}
$$

COROLLARY 16B. *If G^* is a Whitney-dual of G, then G is a Whitney-dual of G^*.*

Proof. This follows immediately from the third equation of theorem 16A.//

Although, as mentioned above, this definition seems highly unintuitive, it turns out to have the properties required of it. The most important of these is stated in the following fundamental theorem due to Whitney.

THEOREM 16C (Whitney 1932). *A graph is planar if and only if it has a Whitney-dual.*

Proof. A direct proof of this result is rather difficult (see Ore[5]), so we shall be content with an indirect one which uses the result of theorem 15F; more precisely, we shall prove that G^* is an abstract-dual of G if and only if G^* is a Whitney-dual of G. Since a graph is planar if and only if it has an abstract-dual, the result will follow immediately.

Suppose then that G^* is an abstract-dual of G. We shall prove that G^* is a Whitney-dual of G by proving that the defining equation for a Whitney-dual remains unchanged if an edge of G is added to the subgraph H; since it is trivially true if H contains no edges, the result will then follow by induction on the number of edges of H. Let us add an edge e to H, and remove the corresponding edge e^* from $\tilde{H}^*$. There are two cases to consider:

(i) If $\gamma(H)$ increases by one when e is added, then the number of components of H must remain unchanged and hence e must join two vertices of H which are already connected by a chain in H; it follows that the addition of the edge e creates a circuit C in G. Since

G^* is an abstract-dual of G, the set C^* of edges of G^* which correspond to the edges in C form a cutset of G^* containing the edge e^*. Hence the number of components of $\tilde{H}^*$ increases by one when e^* is removed, so that $\kappa(\tilde{H}^*)$ decreases by one; the defining equation for a Whitney-dual thus remains unchanged.

(*ii*) If $\gamma(H)$ is unchanged when e is added, then the number of components of H must decrease by one and hence the addition of e creates no new circuit in G. It follows that the removal of e^* creates no new cutset in G^*, and hence that the number of components of $\tilde{H}^*$ must remain unchanged when e^* is removed. We deduce that $\kappa(H^*)$ remains unchanged, and so therefore does the defining equation for a Whitney-dual.

It remains only to show that if G^* is a Whitney-dual of G then G^* is an abstract-dual of G. To prove this, we assume that G^* has n vertices and k components, and that C is a circuit of G. Then $\gamma(C) = 1$ and $\kappa(G^*) = n-k$, and so $\kappa(\tilde{C}^*) = n-k-1$; it follows that C^* is a disconnecting set of G^*. The fact that C^* is actually a cutset of G^* follows from the observation that if E is strictly contained in C, then $\gamma(E) = 0$ and so $\kappa(\tilde{E}^*) = n-k$, which implies that E^* is not a disconnecting set.

The proof of the fact that if C^* is a cutset of G^* then C must be a circuit of G is equally simple, and will be left as an exercise for the reader.//

Exercises

(*16a*) Let G be a plane graph; show directly that if G^* is a geometric-dual of G, then G^* is a Whitney-dual of G.

(**16b*) Let G^* be a Whitney-dual of a connected graph G; show that if G^* has no isolated vertices, then G^* is connected. ★

6 The Colouring of Graphs

With colours fairer
painted their foul ends.

WILLIAM SHAKESPEARE (*The Tempest*)

In this chapter, we investigate the colouring of graphs and maps with special reference to the four-colour conjecture and related topics. We start in §17 by discussing under what conditions the vertices of a graph can be painted in such a way that every edge is incident to vertices of different colours; this discussion spills over into the following section where two major theorems are proved. §19 is devoted to the relationship between the colouring of graphs and the colouring of maps, and both of these are then related in §20 to problems concerning the colouring of the edges of a graph. All of this material is essentially qualitative, asking *whether* graphs can be coloured under certain circumstances, rather than *in how many ways* the colouring can be done; we conclude with a discussion of this second question (using chromatic polynomials) in §21.

§17. THE CHROMATIC NUMBER

If G is a graph without loops, then G is said to be **k-colourable** if to each of its vertices we can assign one of k colours in such a way that no two adjacent vertices have the same colour; if G is k-colourable, but not $(k-1)$-colourable, we say that G is **k-chromatic,** or that the **chromatic number** of G (denoted by $\chi(G)$) is k. Fig. 17.1 shows a graph which is 4-chromatic (and hence k-colourable if $k \geq 4$); the

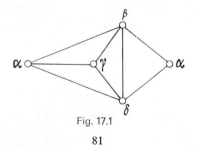

Fig. 17.1

81

colours are denoted by Greek letters. For convenience, we shall assume that *all graphs mentioned in §17 and §18 contain no loops*; we shall however allow multiple edges, since they are irrelevant to our discussion.

It is clear that $\chi(K_n) = n$, and hence we can easily construct graphs with arbitrarily high chromatic number. At the other end of the scale, it is easy to see that $\chi(G) = 1$ if and only if G is a null graph, and that $\chi(G) = 2$ if and only if G is a non-null bipartite graph; it follows from exercise 5g that if G is not a null graph, then $\chi(G) = 2$ if and only if G contains no circuits of odd length. Note in particular that every tree with at least two vertices is 2-chromatic, as is any circuit graph with an even number of vertices.

It is not known under what conditions a graph is 3-chromatic, although it is easy to give examples of such graphs; these examples include the circuit graphs with an odd number of vertices, the wheels with an odd number of vertices, and the Petersen graph. The wheels with an even number of vertices are 4-chromatic.

There is little we can say about the chromatic number of an arbitrary graph; if the graph has n vertices then obviously its chromatic number does not exceed n, and if the graph contains K_r as a subgraph then its chromatic number cannot be less than r, but these results do not take us very far. If however we know the degree of every vertex of the graph, we can usually make significant progress.

THEOREM 17A. *If G is a graph whose largest vertex-degree is ρ, then G is $(\rho+1)$-colourable.*

Proof. The proof is by induction on the number of vertices of G. Let G be a graph with n vertices; then if we remove from G any vertex v and the edges incident to v, the graph which remains is a graph with $n-1$ vertices whose largest vertex-degree is at most ρ. By our induction hypothesis, this graph is $(\rho+1)$-colourable; a $(\rho+1)$-colouring for G is then obtained by colouring v with a different colour from the (at most ρ) vertices adjacent to v.//

By more careful treatment this theorem can be strengthened a little to give the following result which is known as **Brooks' theorem**; its proof will be given in the next section.

THEOREM 17B (Brooks 1941). *If G is a graph whose largest vertex-degree is ρ, then G is ρ-colourable unless (i) G has $K_{\rho+1}$ as a component, or (ii) $\rho = 2$, and G has a circuit of odd length as a component.//*

Both of these theorems are useful if the degrees of all the vertices are approximately equal; for example, we can immediately deduce from theorem 17A that every cubic graph is 4-colourable, and from theorem 17B that every connected cubic graph (apart from K_4) is in fact 3-colourable. On the other hand, if our graph has a few vertices with rather large degrees, then these theorems tell us very little; this is illustrated very well by the star graph $K_{1,n}$ which by Brooks' theorem is n-colourable, but which is in fact 2-chromatic. There is at present no really effective way of avoiding this situation, although the method we shall describe in exercise *17g* helps a little.

This rather depressing situation does not arise if we restrict our attention to planar graphs; in fact we can prove very easily the rather strong result that every planar graph is 6-colourable.

THEOREM 17C. *Every planar graph is 6-colourable.*

Proof. The proof is very similar to that of theorem 17A. We prove the theorem by induction on the number of vertices, the result being trivial for planar graphs with fewer than seven vertices. Suppose then that G is a planar graph with n vertices, and that all planar graphs with $n-1$ vertices are 6-colourable. Without loss of generality G can be assumed to be a simple graph, and so, by theorem 13F, contains a vertex v whose degree is at most five; if we remove v and all the edges which are incident to v, then the graph which remains has $n-1$ vertices and is thus 6-colourable. A 6-colouring for G is then obtained by colouring v with a different colour from the (at most five) vertices adjacent to v.//

As in the case of theorem 17A this result can be made even stronger by more careful treatment, the result being called the **five-colour theorem**; it will also be proved in the next section, but we state it now.

THEOREM 17D. *Every planar graph is 5-colourable.//*

It is natural to ask whether this result can be strengthened further, and this question leads us to the most famous unsolved problem in graph theory—the **four-colour conjecture**; we now state this conjecture formally—an alternative formulation will be given in §**19**.

FOUR-COLOUR CONJECTURE. *Every planar graph is 4-colourable.*

Mathematicians have been trying to prove the four-colour conjecture for about a century, but so far without success. However, significant progress has been made in this direction. We conclude this section by stating without proof some results which have been established; further results will appear later in the chapter.

(*i*) if the four-colour conjecture is false, then any counter-example will be very complicated; for example, it is known that every planar graph with fewer than forty vertices is 4-colourable.

(*ii*) any planar graph which contains no triangles is 3-colourable (Grötzsch's theorem).

(*iii*) if we are trying to prove the four-colour conjecture, then it is sufficient to prove it for Hamiltonian planar graphs (a rather surprising result due to Whitney).

Exercises

(*17a*) Find the chromatic numbers of the Platonic graphs. What can you say about the chromatic numbers of (*i*) the sum of two graphs; (*ii*) the union of two graphs?

(*17b*) Let G be a simple graph with n vertices which is regular of degree d; show that $\chi(G) \geq n/(n-d)$.

(*17c*) Let G be a simple graph whose thickness is t; find an upper bound for the sum of the degrees of all the vertices, and deduce that $\chi(G) \leq 6t$. If, in addition, G has girth r, show that $\chi(G) \leq 2rt/(r-2)$, and deduce that every simple planar graph without triangles is 4-colourable.

(*17d*) A graph is called **critical** if the removal of any vertex (and the edges incident to it) results in a graph of smaller chromatic number; show that (*i*) K_n is critical for all $n > 1$; (*ii*) C_n is critical if and only if n is odd. Show also that if n is odd, then the sum $C_n + C_n$ is a critical graph whose chromatic number is six.

(*17e*) Show that every critical graph which is k-chromatic has the following properties; (*i*) it is connected; (*ii*) every vertex has degree at least $k-1$; (*iii*) there is no vertex whose removal disconnects the graph. Show also that every k-chromatic graph ($k > 1$) contains as a subgraph a k-chromatic graph which is critical, and find such a subgraph for the graph shown in Fig. 17.1.

(*17f*) A conjecture (known as **Hadwiger's conjecture**) asserts that if a connected graph G is k-chromatic, then G is contractible to K_k; prove this conjecture for the cases $k = 2$ and $k = 3$, and show that its truth for the case $k = 5$ implies the truth of the four-colour conjecture.

(**17g*) Let G be a connected simple graph which is not a complete graph or a circuit graph with an odd number of vertices; it can be proved that if λ denotes the largest eigenvalue of G (see exercise *3m*), then $\chi(G) \leq \{\lambda\}$. Use this to obtain upper bounds for the chromatic numbers of (*i*) a star graph; (*ii*) $K_{2,n}$; (*iii*) the cube graph. Compare these upper bounds with those given by Brooks' theorem.

(*17h) Let G be a countable graph, every finite subgraph of which is k-colourable; show that G is k-colourable, and deduce that every countable planar graph is 5-colourable.

(17i) Let χ and $\bar\chi$ denote the chromatic numbers of a simple graph G with n vertices and its complement $\bar G$. Show that $\chi\bar\chi \geq n$, and deduce that $\chi+\bar\chi \geq 2\sqrt n$; use induction on n to show that $\chi+\bar\chi \leq n+1$, and deduce that $\chi\bar\chi \leq \frac{1}{4}(n^2+2n+1)$. Give examples to show that these bounds can all be achieved.

(*17j) How much of the material of this section generalizes to k-graphs?

§18. TWO PROOFS

★In order to avoid disturbing the continuity, we deferred the proofs of theorems 17B and 17D; these proofs will now be given.

THEOREM 17D. *Every planar graph is 5-colourable.*

Proof. The method of proof is similar to that of theorem 17C, although the details are more complicated. We prove the theorem by induction on the number of vertices, the result being trivial for planar graphs with fewer than six vertices. Suppose then that G is a planar graph with n vertices, and that all planar graphs with $n-1$ vertices are 5-colourable. We can assume that G is a simple plane graph, and that (by theorem 13F) G contains a vertex v whose degree is at most five; as before, the removal of v and of all the edges which are incident to v leaves us with a graph with $n-1$ vertices which is thus 5-colourable. Our aim is to colour v in one of our five colours, so completing the 5-colouring of G.

If $\rho(v) < 5$, then v can be coloured with any colour not assumed by the (at most four) vertices adjacent to v, completing the proof in this case. We thus suppose that $\rho(v) = 5$, and that the vertices $v_1, ..., v_5$ which are adjacent to v are arranged around v in clockwise order as in Fig. 18.1. If two of the vertices v_i assume the same

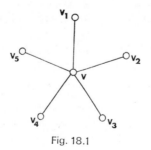

Fig. 18.1

colour, then the proof is complete, since v can then be coloured using a colour which is not assumed by any of the v_i.

We come now to the final case, in which the v_i all assume different colours; let the colour of v_i be c_i ($1 \leq i \leq 5$). Define H_{ij} to be the subgraph of G whose vertices are all those vertices which are coloured c_i or c_j, and whose edges are all those edges which are incident to one vertex of colour c_i and one vertex of colour c_j. There are now two possibilities:

(i) v_1 and v_3 are not in the same component of H_{13} (see Fig. 18.2); in this case, we can interchange the colours of all the vertices in the component of H_{13} containing v_1. The result of this recolouring is that v_1 now has colour c_3, enabling v to be coloured with colour c_1, thus completing the proof in this case.

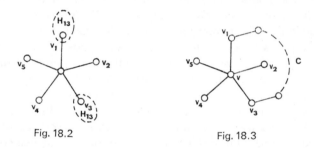

Fig. 18.2   Fig. 18.3

(ii) v_1 and v_3 lie in the same component of H_{13} (see Fig. 18.3); in this case, there is a circuit C of the form $v \to v_1 \to \ldots \to v_3 \to v$, the part between v_1 and v_3 lying entirely in H_{13}. Since v_2 lies inside C and v_4 lies outside C, there cannot be a chain from v_2 to v_4 which lies entirely in H_{24}; we can thus interchange the colours of all the vertices in the component of H_{24} containing v_2. The vertex v_2 now has colour c_4, enabling v to be coloured with colour c_2. This completes the proof.//

We now prove Brooks' theorem; we shall state it in a form slightly different from that given in theorem 17B, although the two statements are easily seen to be equivalent.

THEOREM 18A. *If G is a simple connected graph which is not a complete graph, and whose largest vertex-degree is ρ ($\rho \geq 3$), then G is ρ-colourable.*

Proof. The proof will as usual be by induction on the number of vertices of G. Suppose that G has n vertices; then if any vertex of G has degree less than ρ, the proof may be completed by imitating the proof of theorem 17A. We can thus suppose without loss of generality that G is regular of degree ρ.

We now choose any vertex v and remove it (together with the edges incident to it); the graph which remains is a graph with $n-1$ vertices whose largest vertex-degree is at most ρ. By our induction hypothesis, this graph is ρ-colourable. Our aim is now to colour v in one of the ρ colours; as before, we can suppose that the vertices $v_1, \ldots, v_\rho$ which are adjacent to v are arranged around v in clockwise order, and that they are coloured with distinct colours $c_1, \ldots, c_\rho$.

Defining the subgraphs H_{ij} ($i \neq j$, $1 \leq i, j \leq \rho$) as in the proof of the preceding theorem, we can imitate case (i) of that proof to exclude the case in which v_i and v_j lie in different components of H_{ij}; we may thus assume that given any i and j, v_i and v_j are connected by a chain which lies entirely in H_{ij}. We shall denote the component of H_{ij} which contains v_i and v_j by C_{ij}.

It is clear that if v_i were adjacent to more than one vertex with colour c_j, then there would be a colour (other than c_i) which was not assumed by any of the vertices adjacent to v_i; in this case v_i could be recoloured using this colour, enabling v to be coloured with colour c_i, completing the proof in this case. If this does not happen, then we can use a similar argument to show that every vertex of C_{ij} (other than v_i and v_j) must have degree two; for if w is the first vertex of the chain from v_i to v_j which has degree greater than two, then w can be recoloured using a colour different from c_i or c_j, thereby destroying the property that v_i and v_j are connected by a chain lying entirely in C_{ij}. We can thus assume that for any i, j, the component C_{ij} consists only of a chain from v_i to v_j.

We now remark that two chains of the form C_{ij} and C_{jl} (where $i \neq l$) can be assumed to intersect only at v_j, since if w is another point of intersection, then w can be recoloured using a colour different from c_i, c_j or c_l, contradicting the fact that v_i and v_j are connected by a chain.

To complete the proof, we choose (if possible) two vertices v_i and v_j which are not adjacent, and let w be the vertex with colour c_j which is adjacent to v_i. Since C_{il} is a chain (for any $l \neq j$), we can ˙nterchange the colours of the vertices in this chain without affecting the colouring of the rest of the graph; but this leads to a contradiction, since then w would be a vertex common to the chains C_{ij} and C_{jl}. It follows that it is impossible to choose two vertices v_i and v_j which are not adjacent, and hence G must be the complete graph $K_{\rho+1}$. Since this is excluded, we have covered all possible cases.//

Exercise

(*18a*) Try to prove the four-colour conjecture using the above proof of the five-colour theorem. At what point does the proof fail? ★

§19. THE COLOURING OF MAPS

The four-colour conjecture arose historically in connexion with the colouring of maps. If we have a map (in the non-technical sense!) containing several countries, we may wish to know how many colours are needed to colour the various countries in such a way that no two neighbouring countries share the same colour. Possibly the most familiar form of the four-colour conjecture is the conjecture that every map can be coloured using only four colours.

In order to make this statement more precise, we must say exactly what we mean by a 'map'. Since in the colouring problems we shall be considering, it will be necessary to ensure that the two colours on either side of an edge are different, we shall need to exclude maps in which there is an isthmus. It is convenient, therefore, to define a **map** to be a connected plane graph containing no isthmus. (Note that we do not exclude loops or multiple edges when defining a map; the exclusion of isthmuses corresponds, as we shall see, to the exclusion of loops in §**17**.)

We can now define a map to be **k-colourable(f)** if its faces can be coloured with k colours in such a way that no two adjacent faces (i.e. faces whose boundaries have an edge in common) have the same colour. If there is any possibility of confusion, we shall also use 'k-colourable(v)' to mean k-colourable in the usual sense. As an example, we note that the map shown in Fig. 19.1 is 3-colourable(f) and 4-colourable(v).

The **four-colour conjecture for maps** may now be stated simply as the conjecture that every map is 4-colourable(f). We shall prove the equivalence of the two forms of the four-colour conjecture in corollary 19C. In the meantime, we shall investigate the conditions

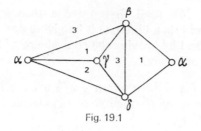

Fig. 19.1

under which a map can be coloured using two colours. It turns out that these conditions take a particularly simple form.

THEOREM 19A. *A map G is 2-colourable(f) if and only if G is an Eulerian graph.*

First proof. ⇒ For any vertex v of G, the faces surrounding v must be even in number since they can be coloured using two colours; it follows that every vertex has even degree and so, by theorem 6B, G is Eulerian.

⇐ We shall describe a method for actually colouring the faces of G. Choose any face F and colour it red; draw a Jordan curve from a point x in F to a point in any other face, making sure that the curve passes through no vertex of G. If the curve from x to a point in face F' crosses an even number of edges, colour F' red; otherwise colour it blue (see Fig. 19.2). The fact that the colouring is well-defined can

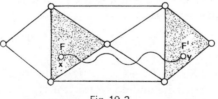

Fig. 19.2

be shown without difficulty by taking a 'circuit' consisting of two such Jordan curves (i.e. a closed Jordan curve) and proving that this circuit crosses an even number of edges of G (using induction on the number of vertices inside the circuit, and the fact that every vertex has an even number of edges incident to it).//

An alternative proof of theorem 19A can, and will, be given by translating the problem into the problem of colouring the vertices of the dual graph. We shall first prove a theorem justifying this procedure, and will then illustrate it by giving our alternative proof of theorem 19A and by proving the equivalence of the two forms of the four-colour conjecture.

THEOREM 19B. *Let G be a planar graph without loops, and let G^* be a geometric-dual of G; then G is k-colourable(v) if and only if G^* is k-colourable(f).*

Proof. ⇒ We may assume that G is plane and connected, so that G^* is a map. If we have a k-colouring(v) for G, then since every face of G^* contains a unique vertex of G, we can k-colour the faces of G^* in such a way that each face inherits the colour of the vertex it contains. The fact that no two adjacent faces of G^* have the same colour follows immediately from the fact that the vertices of G which they contain are adjacent in G and so are differently coloured. Thus G^* is k-colourable(f).

⇐ Suppose now that we have a k-colouring(f) of G^*; then since every vertex of G is contained in a face of G^*, we can k-colour the vertices of G in such a way that each vertex inherits the colour of the face containing it. The fact that no two adjacent vertices of G have the same colour follows immediately by reasoning similar to the above.//

It follows from this result that we can dualize any theorem on the colouring of the vertices of a planar graph to give a theorem on the colouring of the faces of a map, and conversely. As an example of this, consider theorem 19A:

THEOREM 19A. *A map G is 2-colourable(f) if and only if G is an Eulerian graph.*

Second proof. Since (by exercise *15g*) the dual of an Eulerian planar graph is bipartite and conversely, it is sufficient to show that a planar graph without loops is 2-colourable(v) if and only if it is bipartite; but this is obvious.//

We can similarly prove the equivalence of the two forms of the four-colour conjecture.

COROLLARY 19C. *The four-colour conjecture for maps is equivalent to the four-colour conjecture for planar graphs.*

Proof. ⇒ Let G be a planar graph without loops, and assume without loss of generality that G is plane and connected. Then its geometric-dual G^* is a map, and the 4-colourability(v) of G follows immediately from the fact that this map is 4-colourable(f), using theorem 19B.

$\Leftarrow$ Conversely, let G be a map and let G^* be its geometric-dual; then G^* is a planar graph without loops and is therefore 4-colourable(v). It follows immediately that G is 4-colourable(f).//

Duality can also be used to prove the following theorem:

THEOREM 19D. *Let G be a map which is cubic; then G is 3-colourable(f) if and only if every face is bounded by an even number of edges.*

Proof. $\Rightarrow$ Given any face F of G, the faces of G which surround F must alternate in colour; it follows that there must be an even number of them, and hence that every face is bounded by an even number of edges.

$\Leftarrow$ We shall prove the dual result—if G is a connected plane graph without loops, every face of which is a triangle and every vertex of which has even degree (i.e. G is Eulerian), then G is 3-colourable(v). We shall denote the three colours by α, β and γ.

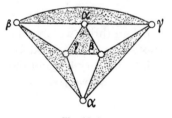

Fig. 19.3

By theorem 19A, since G is Eulerian, the faces of G can be coloured with two colours, say red and blue; the required 3-colouring of the vertices of G is then obtained by first colouring the vertices of any red face, the colouring being such that the colours α, β, and γ appear in clockwise order, and then colouring the vertices of the surrounding faces, the colours α, β and γ appearing in clockwise order around a face if and only if that face is red (see Fig. 19.3). It is easy to see that this colouring of the vertices can be extended to the whole graph, thus proving the theorem.//

In the above theorem, the map was assumed to be cubic; in fact, we can often remove this condition without loss of generality. Our next theorem is a good example of this:

THEOREM 19E. *The four-colour conjecture for planar graphs is true if and only if every cubic map is 4-colourable(f).*

Proof. By corollary 19c, it is sufficient to prove that the 4-colourability(f) of every cubic map implies the 4-colourability(f) of any map.

Let G be any map; then if G contains any vertices of degree two, these vertices can be removed without affecting the colouring. It therefore remains only to show how one can eliminate any vertices of degree four or more; but if v is such a vertex, then we can stick a 'patch' over v (i.e. draw around v a closed Jordan curve which surrounds no vertex except v as in Fig. 19.4). Repeating this for every

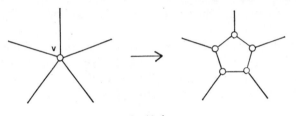

Fig. 19.4

vertex of degree greater than three, we obtain a cubic map which is 4-colourable(f) by hypothesis. The required 4-colouring of the faces of G may then be obtained by shrinking all the patches to zero and reinstating every vertex of degree two.//

Exercises

(*19a*) Show (both directly, and by using duality) that the countries of a map can be coloured with not more than six colours in such a way that neighbouring countries have different colours.

(*19b*) Repeat the previous question with 'six' replaced by 'five'.

(*19c*) The plane is divided into a finite number of regions by drawing infinite straight lines in an arbitrary manner. Show (in three different ways) that these regions can be 2-coloured.

(*19d*) Let G be a simple plane graph with fewer than twelve faces, and suppose that every vertex of G has degree at least three; use exercise *13e* to show that G is 4-colourable (f). Dualize this result.

(*19e*) What can you say about a plane graph which is both 2-colourable (f) and 2-colourable (v)?

(*19f*) Show that if a toroidal graph is embedded on the surface of a torus, then its faces can be coloured using seven colours; show also that there are such graphs which need all seven colours.

(*19g) Show that every graph of genus $g(g \geq 1)$ is h-colourable (f) where $h = [\frac{1}{2}(7+\sqrt{1+48g})]$. (The fact that there are graphs of genus g which need this number was proved in 1968 by Ringel and Youngs. Unfortunately the above result has not been proved for graphs of genus zero!)

§20. EDGE-COLOURINGS

This brief section will be devoted to a study of the colouring of the edges of a graph. It will turn out that the four-colour conjecture for planar graphs is equivalent to a conjecture concerning edge-colourings of cubic maps.

A graph G is said to be **k-colourable(e)** (or **k-edge-colourable**) if its edges can be coloured with k colours in such a way that no two adjacent edges have the same colour; if G is k-colourable(e) but not $(k-1)$-colourable(e), we say that the **edge-chromatic number** (or **chromatic index**) of G is k, and write $\chi_e(G) = k$. Fig. 20.1 shows a graph G for which $\chi_e(G) = 4$.

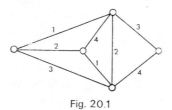

Fig. 20.1

It is clear that if ρ denotes the largest vertex-degree of G, then $\chi_e(G) \geq \rho$. The following result, known as **Vizing's theorem**, gives surprisingly sharp bounds for the edge-chromatic number of G; its proof may be found in Ore.[5]

THEOREM 20A (Vizing 1964). *Let G be a graph without loops, whose largest vertex-degree is ρ; then $\rho \leq \chi_e(G) \leq \rho+1$.//*

It is an unsolved problem to specify exactly which graphs have edge-chromatic number ρ and which have $\rho+1$; however, the results for some particular types of graph can easily be found. For example, $\chi_e(C_n) = 2$ or 3 depending on whether n is even or odd, and $\chi_e(W_n) = n-1$ (if $n \geq 4$); the corresponding results for complete graphs and complete bipartite graphs can also be calculated as we now show.

THEOREM 20B. $\chi_e(K_{m,n}) = \rho = \max (m, n)$.

Remark. It turns out that the edge-chromatic number of any bipartite graph is equal to ρ; the proof of this result will be given in §27.

Proof. We can suppose without loss of generality that $m \geqq n$, and that $K_{m,n}$ is drawn as in Fig. 20.2 with the n vertices in a

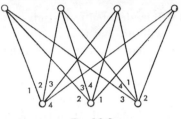

Fig. 20.2

horizontal line below the m vertices. The required edge-colouring is then obtained by successively colouring the edges incident to these n vertices in a clockwise direction using the colours.

$$\{1, 2, ..., m\}; \quad \{2, 3, ..., m, 1\}; \quad ...; \quad \{n, ..., m, 1, ..., n-1\}.//$$

THEOREM 20C. $\chi_e(K_n) = n$ *if n is odd* $(n \neq 1)$, *and* $\chi_e(K_n) = n-1$ *if n is even.*

Proof. If n is odd, then the edges of K_n can be n-coloured by placing the vertices of K_n in the form of a regular n-gon, colouring the edges around the boundary (using a different colour for each edge), and then colouring every remaining edge with the same colour as that used for the boundary edge which is parallel to it (see Fig. 20.3).

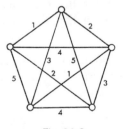

Fig. 20.3

The fact that K_n is not $(n-1)$-colourable(e) follows immediately from the observation that the largest possible number of edges of the same colour is $\frac{1}{2}(n-1)$.

If n is even, then K_n can be regarded as the sum of a complete $(n-1)$-graph K_{n-1} and a single vertex. If the edges of K_{n-1} are then coloured using the method described above, there will be one colour missing at each vertex, and these missing colours will all be different. The colouring of the edges of K_n can thus be completed by colouring the remaining edges with these missing colours (see Fig. 20.4).//

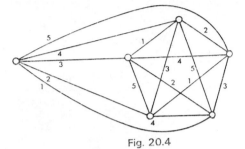

Fig. 20.4

We now show the connexion between the four-colour conjecture and the colouring of the edges of a graph; it is this connexion which accounts for much of the interest in edge-colourings.

THEOREM 20D. *The four-colour conjecture is true if and only if* $\chi_e(G) = 3$ *for every cubic map G.*

Proof. ⇒ Suppose that we are given a 4-colouring of the faces of G, where the colours are denoted by $\alpha = (1, 0), \beta = (0, 1), \gamma = (1, 1)$, and $\delta = (0, 0)$. A 3-colouring of the edges of G can then be obtained by colouring each edge e with the colour obtained by adding together the colours of the two faces adjoining e, this addition being carried out modulo 2; for example, if e adjoins two faces coloured α and γ, then e is coloured β, since $(1, 0) + (1, 1) = (0, 1)$. Note that the colour δ cannot occur in the edge-colouring since the two faces adjoining each edge must be distinct; moreover, it is clearly impossible for any two adjacent edges to share the same colour. We thus have the required edge-colouring (see Fig. 20.5).

⇐ Suppose now that we are given a 3-colouring of the edges of G; then there will be an edge of each colour at each vertex. The subgraph determined by those edges which are coloured α or β is regular of degree two and so the faces of this subgraph can be coloured with two colours which we shall call 0 and 1 (using an obvious extension of theorem 19A to disconnected graphs). In a similar way, the faces of the subgraph determined by those edges which are coloured α or γ can be coloured with the colours 0 and 1. It follows that we can assign to each face of G two coordinates (x, y), where x and y are each 0 or 1. Since the coordinates assigned to two adjacent faces of G must differ in at least one place, it follows that these coordinates $(1, 0)$, $(0, 1)$, $(1, 1)$, $(0, 0)$, give the required 4-colouring of the faces of $G.//$

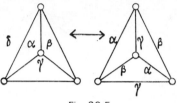

Fig. 20.5

Exercises

(20a)   Calculate the edge-chromatic numbers of the Platonic graphs and the Petersen graph.

(20b)   Show that if G is not a null graph, then the edge-chromatic number of G is the same as the chromatic number of the line graph of G; find all graphs with edge-chromatic number two.

(20c)   Verify theorem 20D in the case of the dodecahedral graph.

(20d)   Let G be a cubic map in which the number of edges surrounding every face is divisible by three; show that G is 4-colourable (f).

(*20e)   Let G be any graph obtained from K_{2n+1} by removing not more than $n-1$ edges; show that $\chi_e(G) = 2n+1$.

§21. CHROMATIC POLYNOMIALS

★We conclude this chapter with a nostalgic glance at vertex-colourings. In this section we shall associate with any graph a function which will tell us, among other things, whether or not the graph is four-colourable; by investigating this function, we may hope to gain some useful information about the four-colour conjecture. Without loss of generality, we shall restrict our attention to simple graphs.

Let G be a simple graph, and let $P_G(k)$ denote the number of ways of colouring the vertices of G with k colours in such a way that no two adjacent vertices have the same colour; P_G will be called (for the time being) the **chromatic function** of G. For example, if G is the graph shown in Fig. 21.1, then $P_G(k) = k(k-1)^2$ since the middle vertex can be coloured in k ways, and consequently the terminal vertices can each be coloured in any of $k-1$ ways; this result can be extended to show that if T is any tree with n vertices, then $P_T(k) = k(k-1)^{n-1}$. Similarly, if G is the complete graph K_3, then $P_G(k) = k(k-1)(k-2)$; this can be extended to give the result $P_G(k) = k(k-1)(k-2) \ldots (k-n+1)$ if G is the graph K_n.

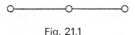

Fig. 21.1

It is clear that if $k < \chi(G)$, then $P_G(k) = 0$, and that if $k \geqq \chi(G)$, then $P_G(k) > 0$. Note also that the four-colour conjecture is equivalent to the statement: if G is a simple planar graph, then $P_G(4) > 0$.

If we are given an arbitrary simple graph, it is difficult in general to obtain the chromatic function by inspection. The following theorem and corollary give us a systematic method for obtaining the chromatic function of a simple graph as the sum of chromatic functions of complete graphs.

THEOREM 21A. *Let G be a simple graph, and v, w be non-adjacent vertices of G. Let G_1 be the graph obtained from G by joining v and w by an edge, and let G_2 be the graph obtained from G by identifying v and w (and identifying multiple edges where necessary); then*

$$P_G(k) = P_{G_1}(k) + P_{G_2}(k).$$

(As an example of this theorem, let G be the graph shown in Fig. 21.2;

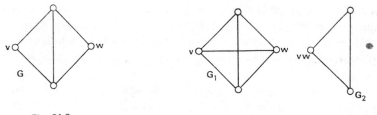

Fig. 21.2   Fig. 21.3

then the corresponding graphs G_1 and G_2 are the graphs shown in Fig. 21.3, and the theorem states that

$$k(k-1)(k-2)^2 = k(k-1)(k-2)(k-3) + k(k-1)(k-2).)$$

Proof. In any allowable colouring of the vertices of G, either v and w have different colours or they have the same colour. The number of colourings in which v and w have different colours is unchanged if an edge is drawn joining v and w, and is therefore equal to $P_{G_1}(k)$. Similarly, the number of colourings in which v and w have the same colour is unchanged if v and w are identified, and is therefore equal to $P_{G_2}(k).//$

COROLLARY 21B. *The chromatic function of a simple graph is a polynomial.*

Proof. The procedure described in the above theorem may be repeated by choosing non-adjacent vertices in G_1 and in G_2 and joining and identifying them in the manner described above, the result being four new graphs. We now repeat the above procedure for these new graphs, and so on. The process terminates when all the pairs of vertices in each graph are adjacent—in other words,

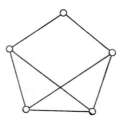

Fig. 21.4

when each graph is a complete graph. Since the chromatic function of a complete graph is a polynomial, it follows by repeated application of theorem 21A that the chromatic function of the graph G must be a sum of polynomials and so must itself be a polynomial. (A worked example to illustrate this will be given later in the section.)$//$

From now on, P_G will be called the **chromatic polynomial** of G. It is easy to see from the proof just given, that if G has n vertices then $P_G(k)$ is of degree n since no new vertices are introduced at any stage. Moreover since the construction yields only one complete graph on n vertices, the coefficient of k^n is one. It can also be shown (see exercise *21d*) that the coefficient of k^{n-1} is $-m$ where m denotes the number of edges of G, and that the coefficients alternate in sign. If there are no colours available, then we cannot colour the graph and it follows that the constant term of the chromatic polynomial must be zero.

It is high time that we gave an example to illustrate the above theory; we shall use theorem 21A to find the chromatic polynomial of the graph G shown in Fig. 21.4 and shall then verify that this polynomial has the form $k^5 - 7k^4 + ak^3 - bk^2 + ck$ (a, b, c positive constants) as the previous paragraph tells us that it must. It is customary at each stage to draw the graph itself, rather than write its chromatic polynomial; for example, instead of writing $P_G(k) = P_{G_1}(k) + P_{G_2}(k)$, where G, G_1 and G_2 denote the graphs of Figs. 21.2 and 21.3, it is convenient to write down the 'equation' given in Fig. 21.5.

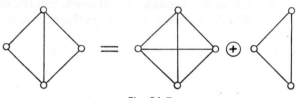

Fig. 21.5

With this convention, we have

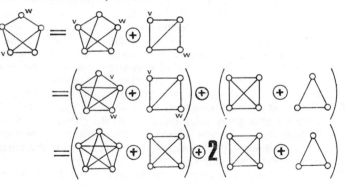

Thus

$$P_G(k) = k(k-1)(k-2)(k-3)(k-4) + 3k(k-1)(k-2)(k-3)$$
$$+ 2k(k-1)(k-2)$$
$$= k(k-1)(k-2)(k^2 - 4k + 5) = k^5 - 7k^4 + 19k^3 - 23k^2 + 10k.$$

(Note that we can immediately deduce that G is 3-chromatic.)

We conclude this chapter with a few remarks to indicate how a study of chromatic polynomials and colourability is related to such subjects as timetabling. Suppose, for example, that we have to arrange the times at which certain lectures are to be given, knowing that some particular lectures cannot be given at the same time (since there may be students who wish to attend both of them); our aim is to find out whether it is possible to construct a timetable which takes account of this. This is done by constructing a graph whose vertices denote the various lectures and whose edges join those pairs of lectures which cannot be scheduled for the same time. If to each time available for lectures we associate a colour, then a colouring of the vertices of the graph corresponds to a successful scheduling of all the lectures, i.e. to a timetable. In this case, a knowledge of the chromatic polynomial of the graph will tell us whether the scheduling is possible, and if so, how many possible ways there are of doing it.

Exercises

(*21a*) Use theorem 21A to find the chromatic polynomial of the octahedral graph.

(*21b*) Find the chromatic polynomials of $K_{1,n}$, $K_{2,n}$, and $K_{3,n}$; can you generalize these results?

(*21c*) What can you say about the chromatic polynomials of (*i*) the union of two simple graphs; (*ii*) the sum of two simple graphs? Show that if G is a disconnected simple graph, then P_G is the product of the chromatic polynomials of its components; in this case, what can you say about the degree of the lowest non-vanishing term?

(**21d*) Let G be a simple graph with n vertices and m edges; prove in detail that (*i*) P_G is a monic polynomial of degree n; (*ii*) the co-efficient of k^{n-1} is $-m$; (*iii*) the coefficients alternate in sign.

(*21e*) Show that G is a tree on n vertices if and only if $P_G(k) = k(k-1)^{n-1}$; deduce that a given polynomial can be the chromatic polynomial of more than one graph. Can you find a polynomial which satisfies the conditions of exercise *21d*, but which is not the chromatic polynomial of any graph?

(**21f*) Let G be a simple graph with m edges, and let $P_G(k)$ be the chromatic polynomial of G; show that the coefficient of k^r in $P_G(k)$ is equal to $\sum_{s=0}^{m} (-1)^s N(r, s)$, where $N(r, s)$ denotes the number of subgraphs of G containing r components and s edges.

(**21g*) How would you define edge-chromatic and face-chromatic polynomials? Investigate their properties. ★

7 Digraphs

*By indirections find
directions out.*
WILLIAM SHAKESPEARE (*Hamlet*)

This chapter and the following one deal with the theory of digraphs and some of its applications. We begin in §22 with the basic definitions, and then discuss under what conditions one can 'direct' the edges of a graph in such a way that the resulting digraph is strongly-connected. This is followed in §23 by a discussion of Eulerian and Hamiltonian dipaths and dicircuits, with particular reference to tournaments. We conclude the chapter with a study of the classification of states of a Markov chain from a digraph point of view.

§22. DEFINITIONS

We begin by recalling some of the definitions of §2. A **digraph** D is defined to be a pair $(V(D), A(D))$, where $V(D)$ is a non-empty finite set of elements called **vertices,** and $A(D)$ is a finite family of ordered pairs of elements of $V(D)$ called **arcs;** $V(D)$ and $A(D)$ are called the **vertex-set** and **arc-family** of D. Thus Fig. 22.1 represents a digraph whose arcs are (u, v), (v, v), (v, w), (v, w), (w, v), (w, u) and (z, w), the ordering of the vertices in an arc being indicated by an arrow. If D is a digraph, the graph obtained from D by 'removing the arrows' (i.e. by replacing each arc of the form (v, w) by a corresponding edge $\{v, w\}$) is called the **underlying graph** of D (see Fig. 22.2).

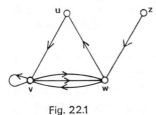

Fig. 22.1

Fig. 22.2

101

We can imitate many of the definitions given in §2 for graphs. For example, two vertices v and w of a digraph D are said to be **adjacent** if there is an arc in $A(D)$ of the form (v, w) or (w, v); the vertices v and w are then said to be **incident** to any such arc. Two digraphs are **isomorphic** if there is an isomorphism between their underlying graphs which preserves the ordering of the vertices in each arc; note in particular that the digraphs shown in Figs. 2.3 and 22.1 are not isomorphic. The **adjacency matrix** of a digraph with vertex-set $\{v_1, ..., v_n\}$ is the matrix $A = (a_{ij})$, in which a_{ij} is the number of arcs in $A(D)$ of the form (v_i, v_j). The matrix in Fig. 22.3 is an adjacency matrix for the digraph of Fig. 22.1. A **simple digraph** is defined in the obvious way.

$$\begin{pmatrix} 0 & 1 & 0 & 0 \\ 0 & 1 & 2 & 0 \\ 1 & 1 & 0 & 0 \\ 0 & 0 & 1 & 0 \end{pmatrix}$$

Fig. 22.3

There are also natural generalizations to digraphs of some of the definitions given in §5. An **arc-sequence** in a digraph D is a finite sequence of arcs of the form $(v_0, v_1), (v_1, v_2), ..., (v_{m-1}, v_m)$. We shall sometimes write this sequence as $v_0 \rightarrow v_1 \rightarrow ... \rightarrow v_m$, and speak of an **arc-sequence from** v_0 **to** v_m. In an analogous way we can define directed paths, directed chains and directed circuits, or (as we shall call them) **dipaths, dichains** and **dicircuits**. Note that although a dipath cannot contain a given arc (v, w) more than once, it can contain both (v, w) and (w, v); for example, in Fig. 22.1, $z \rightarrow w \rightarrow v \rightarrow w \rightarrow u$ is a dipath.

We are now in a position to define connectedness. More precisely, we shall define here the two most natural and useful types of connected digraph, corresponding to whether or not we wish to take account of the direction of the arcs; these definitions are the natural extensions to digraphs of the definitions of connectedness given in §3 and §5.

A digraph D is said to be **connected** (or **weakly-connected**) if it cannot be expressed as the union of two disjoint digraphs (defined in the obvious way); this is equivalent to saying that the underlying graph of D is a connected graph. Suppose, in addition, that for any two vertices v and w of D there is a dichain from v to w; then D is called **strongly-connected**. (This term is so standard that we have used it instead of the more natural 'di-connected'.) It is clear that every strongly-connected digraph is connected, but the converse is not true—Fig. 22.1 shows a connected digraph which is not strongly-connected since there is no dichain from v to z.

The distinction between a connected digraph and a strongly-connected one may become clearer if we consider the road map of a city, all of whose streets are one-way. To say that the road map is connected is to say that we can drive from any part of the city to any other, ignoring the direction of the one-way streets as we go; if the map is strongly-connected, then we can drive from any part of the city to any other, always going the 'right way' down the one-way streets.

It is clearly important that a one-way system should be strongly-connected, and a natural question to ask is, 'when can we impose a one-way system on a street map in such a way that we can drive from any part of the city to any other?' If, for example, the city consists of two parts connected only by a bridge, then we can never impose such a one-way system on the city, since whatever direction we give to the bridge, one part of the city will be cut off. (Note that this includes the case in which we have a cul-de-sac.) If, on the other hand, there are no bridges, then we can always impose such a one-way system; this is the main result of this section and will be stated formally in theorem 22A.

For convenience, we shall define a graph G to be **orientable** if every edge of G (regarded as a pair of vertices) can be ordered in such a way that the resulting digraph is strongly-connected. This process of ordering the edges will be described as 'orienting the graph' or 'directing the edges'. For example, if G is the graph shown in Fig. 22.4, then G can be oriented to give the strongly-connected digraph of Fig. 22.5.

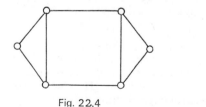

Fig. 22.4

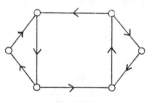

Fig. 22.5

It is easy to see that any Eulerian graph is orientable, since we merely follow any Eulerian path directing the edges in the direction of the path as we go. We now give a necessary and sufficient condition (due to H. E. Robbins) for a graph to be orientable.

THEOREM 22A. *Let G be a connected graph; then G is orientable if and only if each edge of G is contained in at least one circuit.*

Proof. The necessity of the condition is clear. To prove the sufficiency, we choose any circuit C and orient its edges cyclically (in either of the two possible ways). If every edge of G is contained in C, the proof is complete; if not, we choose any edge e which is not in C but which is adjacent to an edge of C. By hypothesis, the edge e is contained in some circuit C' (say) whose edges we may direct cyclically (with the exception of those edges which have already been directed—i.e. those edges of C' which lie also in C). It is not difficult to convince oneself that the resulting digraph is strongly-connected; the situation is illustrated in Fig. 22.6, dashed lines denoting edges

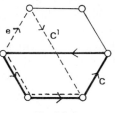

Fig. 22.6

of C'. We proceed in this way, at each stage directing at least one new edge, until the whole graph is oriented. Since, at each stage, the digraph remains strongly-connected, the result follows.//

Exercises

(22a) Let D be a simple digraph with n vertices and m arcs; show that if D is connected but not strongly-connected then
$$n-1 \leq m \leq (n-1)(n-2),$$
and if D is strongly-connected then $n \leq m \leq n(n-1)$.

(22b) Let D be a digraph, and for any two vertices v and w let $v \leq w$ if and only if there is a dichain from v to w; show that $\leq$ is a partial ordering if and only if D contains no dicircuits. How would you characterize a strongly-connected digraph in terms of $\leq$?

(22c)   The **converse** of a digraph D is obtained by reversing the direction of every arc of D; give an example of a digraph which is isomorphic to its converse. What can you say about the adjacency matrices of a digraph and its converse?

(22d)   Let A be the adjacency matrix of a digraph with vertex-set $\{v_1, ..., v_n\}$. Show that the ij-th element of A^k is the number of arc-sequences of length k from v_i to v_j. What meanings can be given to the row-sums and column-sums of A?

(*22e) Three missionaries and three cannibals are on one side of a river which they want to cross; unfortunately, their boat holds only two people, and although all the missionaries can row, only one of the cannibals can. Use the result of the previous exercise to find out how many crossings are needed to transfer everyone to the other side, it being understood that at no time may the cannibals on either bank outnumber the missionaries (unless, of course, there are no missionaries there).

(22f)   How would you define the automorphism group of a digraph? Show that any digraph has the same automorphism group as its converse.

(22g)   Show, without using theorem 22A, that every Hamiltonian graph is orientable. Show, moreover, that $K_n(n \geq 3)$ and $K_{m,n}(m, n \geq 2)$ are orientable, and explicitly find an orientation for each. Find orientations for the Petersen graph and the Platonic graphs.

(*22h) Show how the properties of the symmetric group on three elements (generated by a and b, where $a^3 = b^2 = abab = 1$) can be described using the digraph shown in Fig. 22.7; find similar digraphs describing the symmetric group on four elements and the dihedral groups of orders eight and ten.

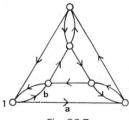

Fig. 22.7

§23. EULERIAN DIGRAPHS AND TOURNAMENTS

In this section we shall attempt to obtain digraph analogues of some of the results of §§6 and 7. This will lead us to the study of Hamiltonian dicircuits in a particular type of digraph called a tournament.

A connected digraph D is called **Eulerian** if there exists a closed

E

dipath which includes every arc of D; such a dipath is called an **Eulerian dipath.** For example, the digraph shown in Fig. 23.1 is not Eulerian, although its underlying graph is an Eulerian graph. Our first aim is to give a necessary and sufficient condition (analogous to the one given in theorem 6B) for a connected digraph to be Eulerian. It is easy to see that one necessary condition is that the digraph is strongly-connected.

We shall need some preliminary definitions. If v is a vertex of a digraph D, we define the **out-degree** of v (denoted by $\overleftarrow{\rho}(v)$, with the arrow 'pointing away from' v) to be the number of arcs of D of the form (v, w); similarly, the **in-degree** of v (denoted by $\overrightarrow{\rho}(v)$) is the

Fig. 23.1

number of arcs of D of the form (w, v). It follows immediately that the sum of the in-degrees of all the vertices of D is equal to the sum of their out-degrees, since each arc of D contributes exactly one to each sum; we shall call this result the **handshaking di-lemma**!

For later convenience, we further define a **source** of D to be a vertex whose in-degree is zero, and a **sink** of D to be one whose out-degree is zero; thus, in Fig. 23.1, v is a source and w is a sink. Note that an Eulerian digraph (other than the trivial one containing no arcs) can contain no sources or sinks.

We are now in a position to state the basic theorem on Eulerian digraphs.

THEOREM 23A. *A connected digraph is Eulerian if and only if* $\overrightarrow{\rho}(v) = \overleftarrow{\rho}(v)$ *for each vertex of D.*

Proof. The proof is entirely analogous to the proof of theorem 6B and will be left as an exercise.$//$

We shall leave the reader to define a semi-Eulerian digraph, and to prove results analogous to corollaries 6C and 6D.

The corresponding study of Hamiltonian dicircuits is, as may be expected, rather less successful than the Eulerian case. A digraph D

is called **Hamiltonian** if there is a dicircuit which includes every vertex of D; a digraph which contains a dichain passing through every vertex is called **semi-Hamiltonian.** Very little is known about Hamiltonian digraphs, and in fact some theorems on Hamiltonian graphs do not seem to generalize easily (if at all) to digraphs. It is natural to ask whether there is a generalization to digraphs of Dirac's theorem (theorem 7A). One such generalization is due to Ghouila–Houri; its proof is considerably more difficult than that of Dirac's theorem, and lies beyond the scope of this book.

THEOREM 23B. *Let D be a strongly-connected digraph with n vertices. If $\vec{\rho}(v) \geqq \frac{1}{2}n$ and $\overleftarrow{\rho}(v) \geqq \frac{1}{2}n$ for each vertex v, then D is Hamiltonian.*||

It seems that results in this direction are not going to come very easily, and so we might consider instead what kinds of digraphs are Hamiltonian. In this respect one type of digraph is very well-known, namely the tournament, the results in this case taking a particularly simple form.

A **tournament** is a digraph in which any two vertices are joined by exactly one arc (see Fig. 23.2). The reason for the name 'tournament'

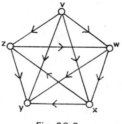

Fig. 23.2

is that the digraph can be used to record the result of a tennis tournament (or any other game in which draws are not allowed). In Fig. 23.2, for example, team z beat team w, but was beaten by team v, etc.

Because of the possibility that a tournament has a source or a sink, tournaments are not in general Hamiltonian. However, the following theorem (due to L. R´dei and P. Camion) shows that every tournament is 'nearly Hamiltonian'.

THEOREM 23C. (*i*) *Every tournament is semi-Hamiltonian*; (*ii*) *every strongly-connected tournament is Hamiltonian.*

Proof. (*i*) The statement is clearly true if the tournament has less than four vertices. We shall prove the result by induction on the number of vertices, and will assume that every tournament on n vertices is semi-Hamiltonian. Let T be a tournament on $n+1$ vertices, and let T' be the tournament on n vertices obtained by removing from T a vertex v and every arc incident to v. Then, by the induction hypothesis, T' has a semi-Hamiltonian dichain $v_1 \rightarrow v_2 \rightarrow \ldots \rightarrow v_n$. There are three cases to consider:

(*1*) if (v, v_1) is an arc in T, then the required dichain is

$$v \rightarrow v_1 \rightarrow v_2 \rightarrow \ldots \rightarrow v_n.$$

(*2*) if (v, v_1) is not an arc in T (which means that (v_1, v) is) and if there exists an i such that (v, v_i) is an arc in T, then choosing i to be

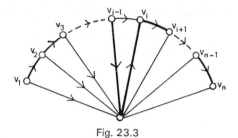

Fig. 23.3

the first such, it is clear that the required dichain is (see Fig. 23.3)

$$v_1 \rightarrow v_2 \rightarrow \ldots \rightarrow v_{i-1} \rightarrow v \rightarrow v_i \rightarrow \ldots \rightarrow v_n.$$

(*3*) if there is no arc in T of the form (v, v_i) then the required dichain is $v_1 \rightarrow v_2 \rightarrow \ldots \rightarrow v_n \rightarrow v$.

(*ii*) We shall prove the stronger result that a strongly-connected tournament T on n vertices contains dicircuits of length 3, 4, ..., n.

To show that T contains a dicircuit of length three, let v be any vertex of T, and let W be the set of all vertices w such that (v, w) is an arc in T, and Z be the set of all vertices z such that (z, v) is an arc. Since T is strongly-connected, W and Z must both be non-empty, and there must be an arc in T of the form (w', z') where w' is in W and z' is in Z. The required circuit of length three is then $v \rightarrow w' \rightarrow z' \rightarrow v$.

It remains only to show that if there is a dicircuit of length k ($k < n$), then there is one of length $k+1$. Let $v_1 \rightarrow \ldots \rightarrow v_k \rightarrow v_1$

be such a dicircuit. Suppose first that there exists a vertex v not contained in this dicircuit, with the property that there exist arcs in T of the form (v, v_i) and of the form (v_j, v). Then there must be a vertex v_i such that both (v_{i-1}, v) and (v, v_i) are arcs in T; the required dicircuit is then

$$v_1 \to v_2 \to \ldots \to v_{i-1} \to v \to v_i \to \ldots \to v_k \to v_1 \text{ (see Fig. 23.4)}.$$

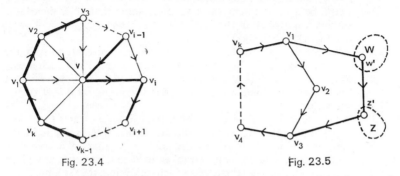

Fig. 23.4   Fig. 23.5

If no vertex exists with the above-mentioned property, then the set of vertices not contained in the dicircuit may be divided into two disjoint sets W and Z, where W is the set of vertices w such that (v_i, w) is an arc for each i, and Z is the set of vertices z such that (z, v_i) is an arc for each i. Since T is strongly-connected, W and Z must both be non-empty, and there must be an arc in T of the form (w', z') where w' is in W and z' is in Z. The required dicircuit is then

$$v_1 \to w' \to z' \to v_3 \to \ldots \to v_k \to v_1 \text{ (see Fig. 23.5).} //$$

Exercises

(23a)   In the tournament shown in Fig. 23.6, find *(i)* dicircuits of lengths three, four and five; *(ii)* an Eulerian dipath; *(iii)* a Hamiltonian dicircuit.

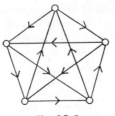

Fig. 23.6

(*23b*) Show that a tournament cannot contain more than one source or one sink.

(*23c*) Find a circular arrangement of nine 1's, nine 2's and nine 3's, in which each of the twenty-seven possible three-digit numbers composed of 1's, 2's and 3's (e.g. 111, 233, etc.) appears exactly once. (Hint: construct the digraph in which the vertices are pairs of integers, and in which there is an arc from ij to kl if and only if $j = k$; now find an Eulerian dipath.)

(*23d*) Let T be a tournament on n vertices; show that if Σ denotes a summation over all of the vertices v of T, then (*i*) $\Sigma \vec{\rho}(v) = \Sigma \overleftarrow{\rho}(v)$; (*ii*) $\Sigma (\vec{\rho}(v))^2 = \Sigma (\overleftarrow{\rho}(v))^2$.

(*23e*) A tournament T is called **irreducible** if it is impossible to split the set of vertices of T into two disjoint sets V_1 and V_2 in such a way that every arc of T is directed from a vertex in V_1 to a vertex in V_2. Show that a tournament is irreducible if and only if it is strongly-connected.

(*23f*) A tournament is called **transitive** if the existence of arcs (u, v) and (v, w) implies the existence of the arc (u, w). Interpret this in the language of tennis tournaments, and show that in a transitive tournament the teams can be ranked in such a way that any given team is better than all the teams which follow it in the ranking.

(*23g*) Use the result of the previous exercise to show that (*i*) a transitive tournament with at least two vertices cannot be strongly-connected; (*ii*) every transitive tournament has a unique semi-Hamiltonian dichain.

(*23h*) The **score** of a vertex of a tournament is its out-degree; the **score-sequence** of a tournament is the sequence formed by arranging the scores of its vertices in non-decreasing order (so that, for example, the score-sequence of the tournament in Fig. 23.2 is (0, 2, 2, 2, 4)). Show that if $(s_1, \ldots, s_n)$ is the score-sequence of a tournament T, then (*i*) $s_1 + \ldots + s_n = \frac{1}{2}n(n-1)$; (*ii*) for any positive integer $k < n$, $s_1 + \ldots + s_k \geqq \frac{1}{2}k(k-1)$, with strict inequality for every value of k if and only if T is strongly-connected; (*iii*) for each k, $\frac{1}{2}(k-1) \leqq s_k \leqq \frac{1}{2}(n+k-2)$; (*iv*) T is transitive if and only if $s_k = k-1$ for each k.

(**23i*) Show that the automorphism group of a tournament must be a group of odd order.

(**23j*) Let T_n be a tournament on n vertices, and let $c(T_n)$ denote the number of dicircuits of length three in T_n. Show that if s_i is the score of the vertex $v_i (1 \leqq i \leqq n)$, then

$$c(T_n) = \binom{n}{3} - \sum_{i=1}^{n} \binom{s_i}{2} = \frac{1}{24}(n^3 - n) - \frac{1}{2} \sum_{i=1}^{n} (s_i - \tfrac{1}{2}(n-1))^2.$$

Deduce that $c(T_n) \leqq \frac{1}{24}(n^3 - n)$ if n is odd, and obtain a corresponding result when n is even.

(*23k) A **bipartite tournament** is a digraph obtained from a complete bipartite graph by giving a direction to each edge. Discuss (with proofs or counter-examples) whether theorem 23c extends to bipartite tournaments.

§24. MARKOV CHAINS

★As the reader may expect, digraphs turn up in a wide variety of 'real-life' situations; rather than trying to cover a large number of these, we shall restrict ourselves to two, referring the reader who is interested in further applications to Chapter 6 of Busacker and Saaty.[2] The two applications we shall present here have been chosen partly for their intrinsic interest and partly because they can be discussed without the introduction of too much preliminary terminology. We start with a not very deep but nonetheless instructive application of digraph theory to the study of Markov chains; the other application—the study of flows in networks—will be discussed in the next chapter.

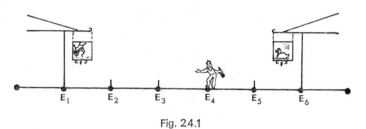

Fig. 24.1

Let us begin by considering the well-known problem of the drunkard who is standing directly between his two favourite pubs, 'The Markov Chain' and 'The Source and Sink' (see Fig. 24.1). Every minute he either staggers ten metres towards the first pub (with probability $\frac{1}{2}$) or towards the second pub (with probability $\frac{1}{3}$) or else he stays where he is (with probability $\frac{1}{6}$)—such a procedure is called a one-dimensional **random walk.** We shall assume also that the two pubs are 'absorbing' in the sense that if he arrives at either of them he stays there. Given the distance between the two pubs and his initial position, there are several questions we can ask; for example, we can ask which pub he is more likely to end up at, and how long he is likely to take getting there.

In order to study the problem of the drunkard in more detail, let us suppose that the two pubs are fifty metres apart and that our friend is initially twenty metres from 'The Source and Sink'. If we denote the various places at which he can stop by $E_1, ..., E_6$, where E_1 and E_6 denote the two pubs, then his initial position E_4 can be described by the vector $x = (0, 0, 0, 1, 0, 0)$, in which the i-th component is the probability that he is initially at E_i. Furthermore, the probabilities of his position after one minute are given by the vector $(0, 0, \frac{1}{2}, \frac{1}{6}, \frac{1}{3}, 0)$, and after two minutes by $(0, \frac{1}{4}, \frac{1}{6}, \frac{13}{36}, \frac{1}{9}, \frac{1}{9})$. It is clearly going to be awkward to calculate directly the probability of his being at a given place after k minutes, and it turns out that the most convenient way of doing this is to introduce the transition matrix.

Let p_{ij} be the probability that he moves from E_i to E_j in one minute; then, for example, $p_{23} = \frac{1}{3}$ and $p_{24} = 0$. These probabilities p_{ij} are called the **transition probabilities,** and the 6×6 matrix $P = (p_{ij})$ is known as the **transition matrix** (see Fig. 24.2);

$$
\begin{pmatrix}
1 & 0 & 0 & 0 & 0 & 0 \\
\frac{1}{2} & \frac{1}{6} & \frac{1}{3} & 0 & 0 & 0 \\
0 & \frac{1}{2} & \frac{1}{6} & \frac{1}{3} & 0 & 0 \\
0 & 0 & \frac{1}{2} & \frac{1}{6} & \frac{1}{3} & 0 \\
0 & 0 & 0 & \frac{1}{2} & \frac{1}{6} & \frac{1}{3} \\
0 & 0 & 0 & 0 & 0 & 1
\end{pmatrix}
$$

Fig. 24.2

note that every entry of P is non-negative and that the sum of the entries in any row is one. It now follows that if x is the initial row vector defined above, then the probabilities of his position after one minute are given by the row vector xP, and after k minutes by the vector xP^k. In other words, the i-th component of xP^k represents the probability that he is at E_i after k minutes have elapsed.

We can generalize these ideas somewhat by defining a **probability vector** to be a row vector whose entries are all non-negative and have unit sum; a **transition matrix** is then defined to be a square matrix, each of whose rows is a probability vector. We can now define a **Markov chain** (or simply, a **chain**) as a pair (P, x) where P is an $n \times n$ transition matrix and x is a $1 \times n$ row vector. If every entry p_{ij} of P is regarded as the (transition) probability of getting from a position E_i to a position E_j, and x is regarded as an initial probability vector, then this definition ties up with the classical definition of a discrete stationary Markov chain to be found in books on probability (see, for example, Feller[14]). The positions E_i are usually referred to as the **states** of the chain, and the aim of this section is to describe various ways of classifying them.

For the remainder of this section we shall be primarily concerned with whether or not we can get from a given state to another state, and if so, what is the shortest time in which this can be done; (for example, in the problem of the drunkard, we can get from E_4 to E_1 in three minutes, but it is impossible to get from E_1 to E_4). It follows that we shall be primarily concerned not with the actual probabilities p_{ij} but with whether or not they are positive, and it is at least reasonable to hope that we may be able to represent the whole set-up

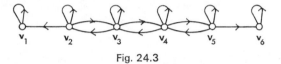

Fig. 24.3

by a digraph in which the vertices correspond to the states and in which the arcs tell us whether we can go from one state to another in one minute. More precisely, if each state E_i is represented by a corresponding vertex v_i then the required digraph is obtained by drawing an arc from v_i to v_j if and only if $p_{ij} \neq 0$; alternatively, the digraph may be defined in terms of its adjacency matrix by replacing each non-zero entry of the matrix P by one. We shall refer to this digraph as the **associated digraph** of the Markov chain; the associated digraph of the one-dimensional random walk is shown in Fig. 24.3. As a further example, if we are given a chain whose transition matrix is the matrix of Fig. 24.4, then its associated digraph is as shown in Fig. 24.5.

It is now clear that we can get from a state E_i to a state E_j in a Markov chain if and only if there is a dichain from v_i to v_j in the associated digraph, and the least possible time taken is then the length of the shortest such dichain. A Markov chain in which we can get from any state to any other is called an **irreducible chain;** clearly a Markov chain is irreducible if and only if its associated digraph is strongly-connected. Note that neither of the chains described above is irreducible.

In investigating these matters further, it is usual to make a distinction between those states to which we keep on returning however long we continue and those which we visit a few times and then never return to. More formally, if on starting at E_i the probability of returning to E_i at some later stage is one, then E_i is called a **persistent** (or **recurrent**) **state**; otherwise E_i is called **transient**. For example, in the problem of the drunkard E_1 and E_6 are trivially persistent, whereas the other states are transient. In more complicated examples, the calculation of the relevant probabilities can become very tricky, and it is often easier to classify the states by analyzing the associated digraph of the chain. It is not difficult to see that a state E_i is persistent if and only if the existence of a dichain from v_i to v_j in the associated digraph implies the existence of a

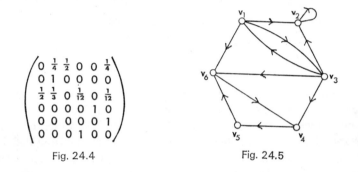

$$\begin{pmatrix} 0 & \frac{1}{4} & \frac{1}{2} & 0 & 0 & \frac{1}{4} \\ 0 & 1 & 0 & 0 & 0 & 0 \\ \frac{1}{2} & \frac{1}{3} & 0 & \frac{1}{12} & 0 & \frac{1}{12} \\ 0 & 0 & 0 & 0 & 1 & 0 \\ 0 & 0 & 0 & 0 & 0 & 1 \\ 0 & 0 & 0 & 1 & 0 & 0 \end{pmatrix}$$

Fig. 24.4   Fig. 24.5

dichain from v_j to v_i. In Fig. 24.5 there is a dichain from v_1 to v_4 but no dichain from v_4 to v_1; it follows that E_1 is transient, and similarly so is E_3 (E_2, E_4, E_5 and E_6 are persistent). A state (such as E_2) from which we can get to no other state is called an **absorbing state**.

An alternative way of classifying states is in terms of their periodicity. A state E_i of a Markov chain is called **periodic of period t** ($t \neq 1$) if it is possible to return to E_i only after a period of time which is a multiple of t; if no such t exists, then E_i is called **aperiodic**. Clearly every state E_i for which $p_{ii} \neq 0$ is aperiodic; it follows that every absorbing state is aperiodic. In the problem of the drunkard the absorbing states E_1 and E_6 are not the only aperiodic states—in fact, every state is aperiodic; on the other hand, in the second example, the absorbing state E_2 is the only aperiodic state since E_1 and E_3 are periodic of period two and E_4, E_5 and E_6 are periodic of period three. In digraph terms, it is easy to see that a state E_i is periodic of period t if and only if in the associated digraph the length of every closed dipath containing v_i is a multiple of t.

Finally, for the sake of completeness, we shall call a state of a Markov chain an **ergodic** state if it is both persistent and aperiodic, and if every state is ergodic then we shall call the chain an **ergodic chain.** For many purposes ergodic chains are the most important and desirable chains to deal with; an example of such a chain will be given in exercise *24e*.

Exercises

(*24a*) Suppose that in the problem of the drunkard, one of the pubs ejects him as soon as he gets there; how will this affect the classification of the states? Will your answer be changed if both pubs eject him?

(*24b*) Make up a problem involving a two-dimensional random walk, and classify the states of the corresponding Markov chain.

(*24c*) Show that if P and Q are transition matrices then so is PQ; what can you say about associated digraphs of P, Q and PQ?

(*24d*) Show how infinite Markov chains may be defined, and construct one in which every state is transient. Show also that every finite Markov chain has at least one persistent state, and deduce that if a finite Markov chain is irreducible then every state is persistent.

(*24e*) A game is being played with a die by n people around a circular table. If the player with the die throws an odd number he passes the die to the player on his left; if he throws a 2 or a 4 he passes it to the player two places to his right; if he throws a 6 he keeps the die and throws again. Show that the resulting Markov chain is ergodic.

(*24f*) What can you say about the states of a Markov chain whose associated digraph is (*i*) Hamiltonian; (*ii*) a tournament? ★

8 Matching, Marriage and Menger's Theorem

They drew all manner of things—
everything that begins with an M—.
LEWIS CARROLL

The results of this chapter are more combinatorial in nature than those of the preceding chapters, although we shall see that they are in fact very closely connected with graph theory. We begin with a discussion of Philip Hall's well-known 'marriage' theorem in several different contexts, including some of its applications to such topics as the construction of latin squares and timetabling problems. This is followed in §28 by a theorem due to Menger on the number of disjoint chains connecting a given pair of vertices in a graph. In §29 we present an alternative formulation of Menger's theorem, known as the max-flow min-cut theorem, which is of fundamental importance in connexion with network flows and transportation problems.

§25. HALL'S 'MARRIAGE' THEOREM

The marriage theorem, proved in 1935 by Philip Hall, answers the following question, known as the **marriage problem**: if we have a finite set of boys each of whom knows several girls, under what conditions can we marry off the boys in such a way that each boy marries a girl he knows? (We shall assume in this section that polygamy is not allowed; the 'general case' will be considered in exercise *25d*.) For example, if there are four boys $\{b_1, b_2, b_3, b_4\}$ and five girls $\{g_1, g_2, g_3, g_4, g_5\}$, and the relationships are as shown in Fig. 25.1, then a possible solution is for b_1 to marry g_4, b_2 to marry g_1, b_3 to marry g_3, and b_4 to marry g_2.

This problem can be represented graphically by taking G to be the bipartite graph in which the vertex-set is divided into two disjoint sets V_1 and V_2 (corresponding to the boys and girls respectively) and in which every edge joins a boy to a girl he knows; Fig. 25.2 shows the graph G corresponding to the situation in Fig. 25.1.

boy	girls known by boy
b_1	g_1 g_4 g_5
b_2	g_1
b_3	g_2 g_3 g_4
b_4	g_2 g_4

Fig. 25.1

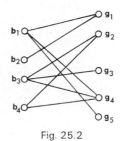

Fig. 25.2

A **complete matching** from V_1 to V_2 in a bipartite graph $G(V_1, V_2)$ is a one-one correspondence between the vertices in V_1 and a subset of the vertices in V_2, with the property that corresponding vertices are joined. It is clear that the marriage problem can be expressed in graph-theoretic terms in the form: 'if $G = G(V_1, V_2)$ is a bipartite graph, when does there exist a complete matching from V_1 to V_2 in G?'

Returning to 'matrimonial terminology', it is clear that a necessary condition for the solution of the marriage problem is that every set of k boys know (collectively) at least k girls (for all integers k satisfying $1 \leq k \leq m$, where m denotes the total number of boys). That this is a necessary condition follows immediately from the fact that if it were not true for a given set of k boys, then we could not marry off the boys in that set, let alone the others.

What is at first sight surprising is that this obviously necessary condition turns out to be sufficient. This is the content of **Hall's 'marriage' theorem**; because of its importance we shall give three proofs, the first of which is due to Halmos and Vaughan.

THEOREM 25A (P. Hall, 1935). *A necessary and sufficient condition for a solution of the marriage problem is that every set of k boys collectively know at least k girls $(1 \leq k \leq m)$.*

Proof. The condition is obviously necessary, as was pointed out above. To prove sufficiency, we shall use induction, and assume that the theorem is true if the number of boys is less than m. (The theorem is clearly true if $m = 1$.) Suppose then that there are m boys; there are two cases to consider:

(*i*) Suppose first that every set of k boys $(1 \leq k < m)$ collectively know at least $k+1$ girls (so that the condition is always true 'with one girl to spare'). Then if we take any boy and marry him to any girl he knows, the original condition remains true for the other $m-1$ boys. These $m-1$ boys can now be married off by induction, completing the proof in this case.

(*ii*) Suppose now that there is a set of k boys ($k < m$) who collectively know exactly k girls. Then these k boys can be married off by induction, leaving $m-k$ boys. But any collection of h of these $m-k$ boys ($1 \leqq h \leqq m-k$) must know at least h of the remaining girls, since otherwise these h boys together with the above collection of k boys would collectively know fewer than $h+k$ girls, contrary to our assumption. It follows that the original condition applies to the $m-k$ boys who can therefore be married off by induction in such a way that everyone is happy and the proof is complete.//

We can also state Hall's theorem in the language of matchings in a bipartite graph; we remind the reader that the number of elements in a set S is denoted by $|S|$.

COROLLARY 25B. *Let* $G = G(V_1, V_2)$ *be a bipartite graph, and for every subset A of* V_1, *let* $\phi(A)$ *be the set of those vertices of* V_2 *which are adjacent to at least one vertex in A; then a complete matching from* V_1 *to* V_2 *exists if and only if* $|A| \leqq |\phi(A)|$ *for each subset A of* V_1.

Proof. The proof of this corollary is simply a translation into graph-theoretic terminology of the above proof.//

Exercises

(*25a*) A building contractor advertises for a bricklayer, a carpenter, a plumber and a toolmaker; he has five applicants—one for the job of bricklayer, one for the job of carpenter, one for the jobs of bricklayer and plumber, and two for the jobs of plumber and toolmaker. Can the jobs be filled? If so, verify in detail the condition given in the statement of Hall's theorem.

(*25b*) Find three more 'real-life' applications of Hall's theorem.

(*25c*) Write out a full proof of corollary 25B.

(*25d*) (The 'harem problem'). Let B be a set of boys, and suppose that each boy in B wishes to marry more than one of his girl friends; find necessary and sufficient conditions for the harem problem to have a solution. (Hint: replace each boy by several identical copies of himself, and then use Hall's theorem).

(*25e*) Show that if every boy has r girl friends and every girl has r boy friends ($r \geqq 1$), then the marriage problem has a solution; deduce that a regular bipartite graph has a complete matching.

(*25f) Suppose that the marriage condition is satisfied, and that each of the m boys knows at least t girls; show, by induction on m, that the marriages can be arranged in at least $t!$ ways if $t \leq m$, and in at least $t!/(t-m)!$ ways if $t > m$.

(*25g) Suppose that the marriage condition is not satisfied; obtain an expression for the maximum number of boys that can be married off to girls known to them.

§26. TRANSVERSAL THEORY

This section is devoted to an alternative proof of Hall's theorem, given in the language of transversal theory; we shall leave the translation of this proof into matching or marriage terminology as an exercise for the reader.

The reader will remember that in our example in the previous section (see Fig. 25.1) the sets of girls known by the four boys were $\{g_1, g_4, g_5\}$, $\{g_1\}$, $\{g_2, g_3, g_4\}$, $\{g_2, g_4\}$ and that a solution of the marriage problem was obtained by finding four distinct g's, one from each of these sets of girls. In general, if E is a non-empty finite set, and $\mathscr{S} = (S_1, ..., S_m)$ is a family of (not necessarily distinct) non-empty subsets of E, then a **transversal** (or **system of distinct representatives**) of $\mathscr{S}$ is a set of m distinct elements of E, one from each set S_i.

To take another example, suppose $E = \{1, 2, 3, 4, 5, 6\}$, and let $S_1 = S_2 = \{1, 2\}$, $S_3 = S_4 = \{2, 3\}$, $S_5 = \{1, 4, 5, 6\}$. Then it is impossible to find five distinct elements of E, one from each subset S_i; in other words, the family $\mathscr{S} = (S_1, ..., S_5)$ has no transversal. Note however that the subfamily $\mathscr{S}' = (S_1, S_2, S_3, S_5)$ has a transversal, for example $\{1, 2, 3, 4\}$. We call a transversal of a subfamily of $\mathscr{S}$ a **partial transversal** of $\mathscr{S}$; in this example $\mathscr{S}$ has several partial transversals (e.g. $\{1, 2, 3, 6\}$, $\{2, 3, 6\}$, $\{1, 5\}$, $\varnothing$, etc.). It is clear that any subset of a partial transversal is a partial transversal.

A natural question to ask is, 'under what conditions does a given family of subsets of a set have a transversal?' The connexion between this problem and the marriage problem is easily seen by taking E to be the set of girls, and S_i to be the set of girls known by boy b_i $(1 \leq i \leq m)$; a transversal in this case is then simply a set of m girls, one corresponding to (and known by) each boy. It follows that theorem 25A gives a necessary and sufficient .condition for a given family of sets to have a transversal; we restate Hall's theorem in this form, and give an alternative proof due to R. Rado.

THEOREM 26A. *Let E be a non-empty finite set, and $\mathscr{S} = (S_1, \ldots, S_m)$ be a family of non-empty subsets of E; then $\mathscr{S}$ has a transversal if and only if the union of any k of the subsets S_i contains at least k elements $(1 \leq k \leq m)$.*

Proof. The necessity of the condition is clear. To prove the sufficiency, we shall show that if one of the subsets $(S_1$, say$)$ contains more than one element, then we can remove an element from S_1 without altering the condition. By repeating this procedure, we can eventually reduce the problem to the case in which each subset contains only one element, the proof then being trivial.

It remains only to show the validity of the 'reduction procedure'. So, suppose that S_1 contains elements x and y, the removal of either of which invalidates the condition. Then there are subsets A and B of $\{2, 3, \ldots, n\}$ with the property that

$$\left| \bigcup_{j \varepsilon A} S_j \cup (S_1 - \{x\}) \right| \leq |A| \quad \text{and} \quad \left| \bigcup_{j \varepsilon B} S_j \cup (S_1 - \{y\}) \right| \leq |B|.$$

But these two inequalities lead to a contradiction, since

$$|A| + |B| + 1 = |A \cup B| + |A \cap B| + 1$$

$$\leq \left| \bigcup_{j \varepsilon A \cup B} S_j \cup S_1 \right| + \left| \bigcup_{j \varepsilon A \cap B} S_j \right| \quad \text{(by the condition)}$$

$$\leq \left| \bigcup_{j \varepsilon A} S_j \cup (S_1 - \{x\}) \right| + \left| \bigcup_{j \varepsilon B} S_j \cup (S_1 - \{y\}) \right| \quad \text{(since } |S_1| \geq 2)$$

$$\leq |A| + |B| \quad \text{(by hypothesis)}.//$$

The beauty of this proof lies in the fact that essentially only one step is involved, in contrast to the Halmos–Vaughan proof which involves the consideration of two separate cases. (It is, however, more awkward to express this proof in the intuitive and appealing language of matrimony!)

Before proceeding to some applications of Hall's theorem, we shall find it convenient to prove two corollaries; these will be needed later (§33).

COROLLARY 26B. *If E and $\mathscr{S}$ are as before, then $\mathscr{S}$ has a partial transversal of size t if and only if the union of any k of the subsets S_i contains at least $k + t - m$ elements.*

Proof. The result follows on applying theorem 26A to the family $\mathscr{S}' = (S_1 \cup D, ..., S_m \cup D)$, where D is any set disjoint from E and containing $m-t$ elements; note that $\mathscr{S}$ has a partial transversal of size t if and only if $\mathscr{S}'$ has a transversal.//

COROLLARY 26C. *If E and $\mathscr{S}$ are as before, and X is any subset of E, then X contains a partial transversal of $\mathscr{S}$ of size t if and only if for every subset A of $\{1, ..., n\}$,*

$$|(\bigcup_{j \in A} S_j) \cap X| \geq |A|+t-m.$$

Proof. The result follows on applying the previous corollary to the family $\mathscr{S}_X = (S_1 \cap X, ..., S_m \cap X)$.//

Exercises

(*26a*) Which of the following families of subsets of $E = \{1, ..., 5\}$ have transversals;

    (*i*) $(\{1\}, \{2, 3\}, \{1, 2\}, \{1, 3\}, \{1, 4, 5\})$;

    (*ii*) $(\{1, 2\}, \{2, 3\}, \{4, 5\}, \{4, 5\})$;

    (*iii*) $(\{1, 3\}, \{2, 3\}, \{1, 2\}, \{3\})$;

    (*iv*) $(\{1, 3, 4\}, \{1, 4, 5\}, \{2, 3, 5\}, \{2, 4, 5\})$?

(*26b*) Let E be the set of letters in the word MATROIDS; how many transversals has the following family of subsets of E: (STAR, ROAD, MOAT, RIOT, RIDS, DAMS, MIST)?

(*26c*) Rewrite the Halmos–Vaughan proof of Hall's theorem in the language of transversal theory; also rewrite Rado's proof in the language of (*i*) matching in a bipartite graph; (*ii*) marriage.

(*26d*) If A is a given subset of E and $\mathscr{S}$ is a family of non-empty subsets of E, show that there exists a transversal of $\mathscr{S}$ which contains A if and only if (*i*) $\mathscr{S}$ has a transversal, and (*ii*) A is a partial transversal of $\mathscr{S}$. (A solution of this exercise using matroid theory will be given in §33).

(**26e*) Let E and $\mathscr{S}$ have their usual meanings; show that if T_1 and T_2 are two transversals of $\mathscr{S}$, and if x is an element of T_1, then there exists an element y of T_2 with the property that $(T_1 - \{x\}) \cup \{y\}$ (the set obtained from T_1 on replacing x by y) is also a transversal of $\mathscr{S}$. Can you see any connexion between this result and that of exercise *9g*?

(**26f*) The **rank** $\rho(A)$ of a subset A of E is defined as the number of elements in the largest partial transversal of $\mathscr{S}$ contained in A; show that (*i*) $0 \leq \rho(A) \leq |A|$; (*ii*) if $A \subseteq B \subseteq E$, then $\rho(A) \leq \rho(B)$; (*iii*) for any A, $B \subseteq E$, $\rho(A \cup B)+\rho(A \cap B) \leq \rho(A)+\rho(B)$. Do these results seem at all familiar?

(*26g) Let E be a countable set, and let $\mathscr{S} = (S_1, S_2, \ldots)$ be a countable family of non-empty *finite* subsets of E. Defining a transversal of $\mathscr{S}$ in the natural way, show (by using König's lemma) that $\mathscr{S}$ has a transversal if and only if the union of any k of the subsets S_i contains at least k elements (for all finite k). By considering the example $E = \{1, 2, 3, \ldots\}$, $S_1 = \{E\}$, $S_2 = \{1\}$, $S_3 = \{2\}$, $S_4 = \{3\}$, ..., show that this result ceases to be true if not all of the S_i are finite.

§27. APPLICATIONS OF HALL'S THEOREM

★In this section we shall apply Hall's theorem to four different fields, namely (*i*) the construction of latin squares; (*ii*) a problem concerning (0, 1)-matrices; (*iii*) the colouring of the edges of a bipartite graph, and (*iv*) the existence of a common transversal of two families of subsets of a set, and its relevance to timetabling problems.

(*i*) LATIN SQUARES. An $m \times n$ **latin rectangle** is an $m \times n$ matrix $M = (m_{ij})$ whose entries are integers satisfying (*1*) $1 \leqq m_{ij} \leqq n$, (*2*) no two entries in any row or in any column are equal. Note that (*1*) and (*2*) imply that $m \leqq n$; if $m = n$, then the latin rectangle is called a **latin square**. For example, Figs. 27.1 and 27.2 show a 3×5

$$\begin{pmatrix} 1 & 2 & 3 & 4 & 5 \\ 2 & 4 & 1 & 5 & 3 \\ 3 & 5 & 2 & 1 & 4 \end{pmatrix}$$

$$\begin{pmatrix} 1 & 2 & 3 & 4 & 5 \\ 2 & 4 & 1 & 5 & 3 \\ 3 & 5 & 2 & 1 & 4 \\ 4 & 3 & 5 & 2 & 1 \\ 5 & 1 & 4 & 3 & 2 \end{pmatrix}$$

Fig. 27.1   Fig. 27.2

latin rectangle and a 5×5 latin square. We can ask the following question: given an $m \times n$ latin rectangle with $m < n$, when can we adjoin $n - m$ new rows in such a way that a latin square is produced? Surprisingly, the answer is 'always'!

THEOREM 27A. *Let M be an $m \times n$ latin rectangle with $m < n$; then M can be extended to a latin square by the addition of $n - m$ new rows.*

Proof. We shall prove that M can be extended to an $(m+1) \times n$ latin rectangle; by repeating the procedure involved, we eventually obtain a latin square.

Let $E = \{1, 2, ..., n\}$, and $\mathscr{S} = (S_1, ..., S_n)$, where S_i denotes the set consisting of those elements of E which do *not* occur in the i-th column of M. If we can prove that $\mathscr{S}$ has a transversal then the proof is complete, since the elements in this transversal will form the additional row. By Hall's theorem, it is sufficient to show that the union of any k of the S_i contains at least k distinct elements; but this is obvious, since such a union contains $(n-m)k$ elements altogether (including repetitions), and if there were fewer than k distinct elements then at least one of them would have to appear more than $n-m$ times, which is a contradiction.//

(*ii*) (0, 1)-MATRICES. An alternative way of studying transversals of a family $\mathscr{S} = (S_1, ..., S_m)$ of non-empty subsets of a set $E = \{e_1, ..., e_n\}$ is to study the **incidence matrix** of the family, i.e. the $m \times n$ matrix $A = (a_{ij})$ in which $a_{ij} = 1$ if $e_j \, \varepsilon \, S_i$, and $a_{ij} = 0$ otherwise. (We shall call such a matrix, in which every entry is 0 or 1, a **(0, 1)-matrix**.) If we define the **term rank** of A to be the largest number of 1's of A, no two of which lie in the same row or column, then $\mathscr{S}$ has a transversal if and only if the term rank of A is m; moreover, the term rank of A is precisely the number of elements in a partial transversal of largest possible size. We now prove, as a second application of Hall's theorem, a famous result on (0, 1)-matrices known as the **König-Egerváry theorem.**

THEOREM 27B (König–Egerváry 1931). *The term rank of a* (0, 1)-*matrix A is equal to the minimum number μ of rows and columns which together contain all the* 1's *of A.*

Remark. As an illustration of the theorem, consider the matrix of Fig. 27.3 which is the incidence matrix of the second family $\mathscr{S}$ described on page 119; clearly the term rank and μ are both four.

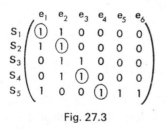

Fig. 27.3

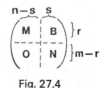

Fig. 27.4

Proof. It is obvious that the term rank cannot exceed μ. To prove equality, we can suppose without loss of generality that all of the 1's of A are contained in r rows and s columns (where $r+s = \mu$), and that the order of the rows and columns is such that A contains, in the bottom left-hand corner, an $(m - r) \times (n - s)$ submatrix consisting entirely of zeros (Fig. 27.4). If $i \leq r$, define S_i to be the set of integers $j \leq n - s$ such that $a_{ij} = 1$. It is a straightforward exercise to check that the union of any k of the S_i contains at least k integers, and hence that the family $\mathscr{S} = (S_1,...,S_r)$ has a transversal. It follows that the submatrix M of A contains a set of r 1's, no two of which lie in the same row or column; similarly, the matrix N contains a set of s 1's with the same property. Hence A contains a set of $r+s$ 1's, no two of which lie in the same row or column, showing that μ cannot exceed the term rank, as required.//

We have just proved the König–Egerváry theorem, using Hall's theorem; it is even easier to prove Hall's theorem using the König–Egerváry theorem (see exercise *27c*). It follows that the two theorems are, in some sense, equivalent. Later on in this chapter we shall be proving Menger's theorem and the max-flow min-cut theorem, both of which can also be shown to be equivalent to Hall's theorem.

(*iii*) EDGE-COLOURINGS OF GRAPHS. In §**20** we showed that if G is a complete bipartite graph whose largest vertex-degree is ρ, then the edge-chromatic number of G is equal to ρ. We shall now use a result from this chapter to show that this result holds also for an arbitrary bipartite graph.

THEOREM 27C. If G is a bipartite graph whose largest vertex-degree is ρ, then $\chi_e(G) = \rho$.

Proof. By the result of exercise *3g*, there exists a bipartite graph $G' = G'(V_1, V_2)$ which is regular of degree ρ, which contains G as a subgraph, and in which $|V_1| = |V_2|$. By exercise *25e*, G' has a complete matching; note that this matching includes every vertex of G', and that any vertex of G of degree ρ corresponds under the matching to another vertex of G.

To prove that G has edge-chromatic number ρ, it is sufficient to take this matching of G' and colour with the first colour those edges of G which are incident to a vertex of G of degree ρ. The rest of G will then be a bipartite graph whose largest vertex-degree is $\rho-1$, and we can use induction to assert that its edges can be coloured using the remaining $\rho - 1$ colours, thus completing the proof.//

(*iv*) COMMON TRANSVERSALS. We conclude this section with a brief discussion of common transversals. If E is a non-empty finite set and $\mathscr{S} = (S_1, ..., S_m)$ and $\mathscr{T} = (T_1, ..., T_m)$ are two families of non-empty subsets of E, it is of interest to know when there exists a **common transversal** for $\mathscr{S}$ and $\mathscr{T}$, i.e. a set of m distinct elements of E which form a transversal of both $\mathscr{S}$ and $\mathscr{T}$. In timetabling problems, for example, if E denotes the set of times at which lectures may be given, the sets S_i denote the times that m given professors are willing to lecture, and the sets T_i denote the times that m lecture rooms are available, then the finding of a common transversal of $\mathscr{S}$ and $\mathscr{T}$ enables us to assign each professor to an available lecture room at a time suitable to him.

We can in fact give a necessary and sufficient condition for two families to have a common transversal; note that theorem 27D reduces to Hall's theorem if we put $T_j = E$ for $1 \leq j \leq m$.

THEOREM 27D. *Let E be a non-empty finite set, and let $\mathscr{S} = (S_1, ..., S_m)$ and $\mathscr{T} = (T_1, ..., T_m)$ be two families of non-empty subsets of E; then $\mathscr{S}$ and $\mathscr{T}$ have a common transversal if and only if, for all subsets A and B of $\{1, 2, ..., m\}$,*

$$|(\bigcup_{i \varepsilon A} S_i) \cap (\bigcup_{j \varepsilon B} T_j)| \geq |A| + |B| - m.$$

Sketch of proof. Consider the family $\mathscr{U} = \{U_i\}$ of subsets of $E \cup \{1, ..., m\}$ (assuming E and $\{1, ..., m\}$ to be disjoint), where the indexing set is also $E \cup \{1, ..., m\}$ and where $U_i = S_i$ if $i \varepsilon \{1, ..., m\}$ and $U_i = \{i\} \cup \{j : i \varepsilon T_j\}$ if $i \varepsilon E$.

It is not difficult to verify that $\mathscr{S}$ and $\mathscr{T}$ have a common transversal if and only if $\mathscr{U}$ has a transversal. The result then follows on applying Hall's theorem to the family $\mathscr{U}$.//

At the time of writing, it is not known under what conditions there exists a common transversal for three families of non-empty subsets of a set, and the problem of finding such conditions seems to be very difficult. Many attempts to solve this problem use matroid theory; in fact, as we shall see in the next chapter, several problems in transversal theory (for example, exercise *26d* and theorem 27D) become almost trivial when looked at from this viewpoint. Further results in transversal theory may also be found in Mirsky.[10]

Exercises

(27a) Show how the multiplication table of a group can be regarded as a latin square; give an example of a latin square which cannot be obtained in this way.

(27b) Show that there are at least $n!(n-1)!\ldots 1!$ $n \times n$ latin squares; obtain a corresponding lower bound for the number of $m \times n$ latin rectangles.

(27c) Show how one can use the König–Egerváry theorem to deduce Hall's theorem.

(27d) Use theorem 27D to show that if G is a finite group, H is a subgroup of G, and $G = x_1H \cup x_2H \cup \ldots \cup x_mH = Hy_1 \cup Hy_2 \cup \ldots \cup Hy_m$ are left and right coset decompositions of G with respect to H, then there exist elements $z_1, \ldots, z_m$ in G with the property that
$$G = z_1H \cup z_2H \cup \ldots \cup z_mH = Hz_1 \cup Hz_2 \cup \ldots \cup Hz_m.$$

(*27e) Show that if A is a (0, 1)-matrix in which the sum of the elements in each row or column is k, then A can be expressed as the sum of k matrices each of which contains exactly one 1 in each row and column; deduce the first part of exercise 25e as a corollary. ★

§28. MENGER'S THEOREM

We now discuss a theorem which turns out to be closely related to Hall's theorem, and which has very far-reaching practical applications. This theorem is due to Menger and concerns the number of chains connecting two given vertices v and w in a graph G. We might need, for example, to find the maximum number of chains from v to w, no two of which have an edge in common—such chains are called **edge-disjoint chains**; alternatively, we may want to find the maximum number of chains from v to w, no two of which have a vertex in common (except, of course, v and w)—these are called **vertex-disjoint chains**. (In the graph of Fig. 28.1, there are clearly four edge-

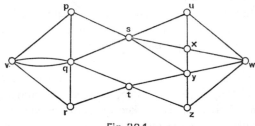

Fig. 28.1

disjoint chains and two vertex-disjoint ones.) Analogously, we can ask for the maximum number of vertex-disjoint or arc-disjoint dichains (defined in the obvious way) connecting two vertices v and w in a digraph; in this case we can, without loss of generality, take v to be a source and w to be a sink. We shall be concentrating primarily on graphs, the corresponding discussion for digraphs being left to the reader.

In order to investigate these problems, we shall need some further definitions; we shall assume throughout that G is a connected graph and that v and w are given distinct vertices of G. A **vw-disconnecting set** of G is a set E of edges of G with the property that any chain from v to w includes an edge of E; note that a vw-disconnecting set is a disconnecting set of G. Similarly, a **vw-separating set** of G is a set S of vertices (not including v or w) with the property that any chain from v to w passes through a vertex of S. In Fig. 28.1, for example, the sets $E_1 = \{\{p, s\}, \{q, s\}, \{t, y\}, \{t, z\}\}$ and $E_2 = \{\{u, w\}, \{x, w\}, \{y, w\}, \{z, w\}\}$ are vw-disconnecting sets, and $V_1 = \{s, t\}$ and $V_2 = \{p, q, y, z\}$ are vw-separating sets.

In order to count the number of edge-disjoint chains from v to w, we first note that if E is a vw-disconnecting set containing k edges, then the number of edge-disjoint chains cannot possibly exceed k, since otherwise some edge in E would be included in more than one chain. If, moreover, E is a vw-disconnecting set of smallest possible size, then it turns out that the number of edge-disjoint chains is actually equal to k, and that consequently there is exactly one edge of E in each such chain. This result is known as the edge-form of **Menger's theorem,** although it was in fact first proved by Ford and Fulkerson in 1955.

THEOREM 28A. *The maximum number of edge-disjoint chains connecting two distinct vertices v and w of a connected graph G is equal to the minimum number, k, of edges in a vw-disconnecting set.*

Remark. The proof we are about to give is non-constructive, in the sense that if we are given G, it will not provide us with a systematic way of obtaining k edge-disjoint chains, or even of finding the value of k; an algorithm which can be used to solve these problems will be given in the next section.

Proof. As we have just pointed out, the maximum number of edge-disjoint chains connecting v and w cannot exceed the minimum number of edges in a vw-disconnecting set. We shall use induction on the number of edges of G to prove that these numbers are actually equal. Suppose that the number of edges of G is m, and that the theorem is true for all graphs with fewer than m edges. There are two cases to consider:

(*i*) We suppose first that there exists a vw-disconnecting set E of minimum size k, with the property that not all of its edges are incident to v, and not all of them are incident to w; for example, in the graph of Fig. 28.1, the set E_1 defined above would be such a vw-disconnecting set. The removal from G of the edges in E leaves two disjoint subgraphs V and W containing v and w respectively. We now define two new graphs G_1 and G_2 as follows: G_1 is obtained from G by contracting every edge of V (i.e. by shrinking V down to v) and G_2 is obtained similarly by contracting every edge of W. (The graphs G_1 and G_2 obtained from Fig. 28.1 are shown in Fig. 28.2;

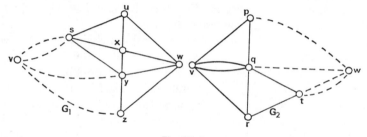

Fig. 28.2

the dashed lines denote edges of E_1). Since G_1 and G_2 have fewer edges than G and since E is clearly a vw-disconnecting set of minimum size for both G_1 and G_2, the induction hypothesis tells us that there are k edge-disjoint chains in G_1 from v to w, and similarly for G_2. The required k edge-disjoint chains in G are then obtained by combining these chains in the obvious way.

(*ii*) We now suppose that every vw-disconnecting set of minimum size k consists only of edges which are all incident to v or all incident to w; for example, in Fig. 28.1, the set E_2 is such a vw-disconnecting set. We can assume without loss of generality that every edge of G is contained in a vw-disconnecting set of size k, since otherwise its removal would not affect the value of k and we could use the induction hypothesis to obtain k edge-disjoint chains. It follows that if C

is any chain from v to w, then C must consist either of a single edge or of two edges, and can thus contain at most one edge of any vw-disconnecting set of size k. By removing from G the edges of C, we obtain a graph which contains at least $k-1$ edge-disjoint chains (by the induction hypothesis); these chains, together with C, give the required k chains in G.//

We turn now to the other problem mentioned at the beginning of the section—namely, to find the number of vertex-disjoint chains from v to w. (It was actually this problem which Menger himself solved, although his name is usually given to both theorem 28A and theorem 28B.) What is at first sight rather surprising is that not only does its solution have a form very similar to theorem 28A, but also the proof of theorem 28A goes through with only minor changes, mainly involving the replacement of such terms as 'edge-disjoint' and 'incident' by 'vertex-disjoint' and 'adjacent'. We now state the vertex-form of Menger's theorem—its proof will be left to the reader.

THEOREM 28B (Menger 1927). *The maximum number of vertex-disjoint chains connecting two distinct non-adjacent vertices v and w of a graph G is equal to the minimum number of vertices in a vw-separating set.*//

As we pointed our earlier, the above discussion can be modified to give the number of vertex-disjoint or arc-disjoint dichains in a digraph in terms of disconnecting sets and separating sets. In this case, a vw-disconnecting set is a set A of arcs with the property that every dichain from v to w includes an arc in A. Once again the corresponding theorem takes a form very similar to theorem 28A, and the proof goes through almost word for word. We state it formally as the **integrity theorem** (the reason for this name will become apparent in the following section).

THEOREM 28C (Integrity theorem). *The maximum number of arc-disjoint dichains connecting two distinct vertices v and w of a digraph D is equal to the minimum number of arcs in a vw-disconnecting set.*//

As an example of the integrity theorem, we let D be the digraph shown in Fig. 28.3. It is straightforward to verify that there are six arc-disjoint dichains from v to w; a corresponding vw-disconnecting set is indicated by dashed lines.

As the reader can see, these diagrams are likely to become very cumbersome as the number of arcs joining pairs of adjacent vertices

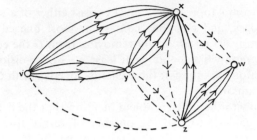

Fig. 28.3

increases; this can be overcome by drawing just one arc and writing next to it the number of arcs there should be (see Fig. 28.4). This seemingly innocent remark turns out to be fundamental in the study of network flows and transportation problems, which will be discussed in the following section.

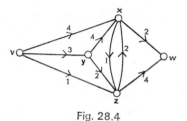

Fig. 28.4

We end this section by proving that Hall's theorem can be deduced from Menger's theorem. We shall prove the version of Hall's theorem that appears in corollary 25B.

THEOREM 28D. *Menger's theorem implies Hall's theorem.*

Proof. Let $G = G(V_1, V_2)$ be a bipartite graph; we have to prove that if $|A| \leq |\phi(A)|$ for each subset A of V_1 (using the notation of corollary 25B), then there exists a complete matching from V_1 to V_2. This is done by applying the vertex-form of Menger's theorem (theorem 28B) to the graph obtained by adjoining to G a vertex v adjacent to every vertex in V_1 and a vertex w adjacent to every vertex in V_2 (see Fig. 28.5). Since a complete matching from V_1 to V_2 exists if and only if the number of vertex-disjoint chains from v to w is equal to the number of vertices in V_1 ($=k$, say), it is enough to show that every vw-separating set contains at least k vertices.

Let S be a vw-separating set, consisting of a subset A of V_1 and a subset B of V_2. Since $A \cup B$ is a vw-separating set, there can be no edges joining a vertex of $V_1 - A$ to a vertex of $V_2 - B$, and hence $\phi(V_1 - A) \subseteq B$. It follows that $|V_1 - A| \leq |\phi(V_1 - A)| \leq |B|$, and so $|S| = |A| + |B| \geq |V_1| = k$, as required.//

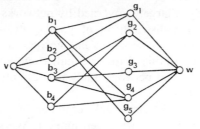

Fig. 28.5

Exercises

(*28a*) Verify both the edge-form and the vertex-form of Menger's theorem for the Petersen graph (for all possible choices of the vertices v and w).

(*28b*) Prove theorem 28B in detail, and verify it for the cube graph shown in Fig. 28.6.

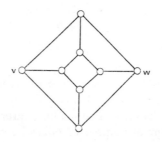

Fig. 28.6

(*28c*) Show how the vertex-form and edge-form of Menger's theorem may be deduced from each other.

(*28d*) A graph G is said to be **k-connected** if every pair of non-adjacent vertices are connected by at least k vertex-disjoint chains. Show that (*i*) $W_n(n \geq 4)$ is 3-connected; (*ii*) $K_{m,n}$ is k-connected, where $k = \min(m, n)$; (*iii*) if G is k-connected, then the degree of every vertex is at least k; (*iv*) G is 2-connected if and only if every pair of vertices of G are contained in a circuit; (*v*) G is k-connected if and only if k is the smallest number of vertices which must be removed in order to disconnect G.

(*28e*) Show how theorems 27B and 27D can both be deduced from Menger's theorem.

(*28f*) Show how Menger's theorem can be extended to infinite graphs.

§29. NETWORK FLOWS

Our society today is largely governed by networks—transportation, communication, the distribution of goods, etc.—and the mathematical analysis of such networks has become a subject of fundamental importance. In this section we shall attempt to show by means of simple examples that network analysis is essentially equivalent to the study of digraphs.

A manufacturer of electric back-scratchers wants to send several boxes of back-scratchers to a given market. We shall assume that there are various channels through which the boxes can be sent, and that these channels are as shown in Fig. 29.1 (with v representing the

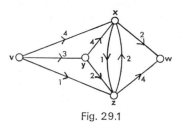

Fig. 29.1

manufacturer and w the market); the numbers appearing on the diagram refer to the maximum loads which may be passed through the corresponding channels. The manufacturer is clearly interested in finding the maximum number of boxes he can send through the network without exceeding the permitted capacity of any channel.

Fig. 29.1 can also be used to describe other situations. For example, if each arc of the digraph represents a one-way street and the number associated with each street refers to the maximum possible flow of traffic (in vehicles per hour) along that street, then we may want to find the greatest possible number of vehicles which can travel from v to w in one hour. Alternatively we can regard the diagram as depicting an electrical network, the problem then being to find the maximum current which can safely be passed through the network given the currents at which the individual wires burn out.

Using these examples as motivation, we may now define a
network N to be a digraph to each arc a of which has been assigned
a non-negative real number $\psi(a)$ called its **capacity**; equivalently, a
network may be defined as a pair (D, ψ) where D is a digraph and ψ
is a function from the arc-set of D to the set of non-negative real
numbers. The **out-degree** $\overleftarrow{\rho}(x)$ of a vertex x is then defined to be the
sum of the capacities of the arcs of the form (x, z), and the **in-
degree** $\overrightarrow{\rho}(x)$ is similarly defined. For example, in the network of
Fig. 29.1, $\overleftarrow{\rho}(v) = 8$ and $\overrightarrow{\rho}(x) = 10$. It is clear that the analogue of
the handshaking di-lemma then takes the following form: the sum
of the out-degrees of the vertices of a network is equal to the sum
of the in-degrees. In the following, we shall always assume (unless

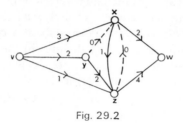

Fig. 29.2

otherwise stated) that the digraph D contains exactly one source v
and one sink w; the general case of several sources and sinks (corresponding in the first example above to more than one manufacturer
and market) may be easily reduced to this special case (see exercise
29b).

Given a network $N = (D, \psi)$, we define a **flow in N** to be a
function ϕ which assigns to each arc a of D a non-negative real
number $\phi(a)$ (called the **flow in a**) in such a way that (i) for any arc
a, $\phi(a) \leqq \psi(a)$; (ii) with respect to the network (D, ϕ), the out-degree
and in-degree of any vertex (other than v or w) are equal. Informally
this means that the flow in any arc cannot exceed its capacity, and
that the 'total flow' into any vertex (other than v or w) is equal to the
'total flow' out of it. Fig. 29.2 gives a possible flow for the network
of Fig. 29.1; another flow is the **zero flow** in which the flow in every
arc is zero (any other flow being called a non-zero flow). For
convenience, we shall say that an arc a for which $\phi(a) = \psi(a)$ is
called **saturated**; in Fig. 29.2, the arcs (v, z), (x, z), (y, z), (x, w) and
(z, w) are saturated, the remaining arcs being called **unsaturated.**

It follows from the handshaking di-lemma that the sum of the flows in the arcs incident to v is equal to the sum of the flows in the arcs incident to w; this sum is called the **value of the flow.** Prompted by the examples considered at the beginning of this section, we shall be primarily interested in those flows whose value is as large as possible—the so-called **maximal flows;** the reader can easily check that the flow of Fig. 29.2 is a maximal flow for the network of Fig. 29.1, and that its value is six. Note that in general a network can have several different maximal flows but that their values must all be equal.

The study of maximal flows in a network $N = (D, \psi)$ is closely tied up with the concept of a **cut,** which is simply a set A of arcs of D with the property that every dichain from v to w includes an arc in A; in other words, a cut in a network is merely a vw-disconnecting set in the corresponding digraph D. The **capacity of a cut** is then defined to be the sum of the capacities of the arcs in the cut. We shall be concerned mainly with those cuts whose capacity is as small as possible, the so-called **minimal cuts;** in Fig. 29.1, an example of a minimal cut is provided by the arcs (v, z), (x, z), (y, z) and (x, w), the capacity of this cut being six.

It is clear that the value of any flow cannot exceed the capacity of any cut, and hence that the value of any maximal flow cannot exceed the capacity of any minimal cut. What is not immediately clear is that these last two numbers are always equal; this famous result is known as the **max-flow min-cut theorem** and was first proved by Ford and Fulkerson in 1955. We shall present two proofs; the first one shows that the max-flow min-cut theorem is essentially equivalent to Menger's theorem, whereas the second one is a direct proof.

THEOREM 29A (Max-flow min-cut theorem). *In any network, the value of any maximal flow is equal to the capacity of any minimal cut.*

First proof. We shall suppose to begin with, that the capacity of every arc is an integer. In this case, the network can be regarded as a digraph $\tilde{D}$ in which the capacities represent the number of arcs connecting the various vertices (see Figs. 28.3 and 28.4). The value of a maximal flow then corresponds to the total number of arc-disjoint dichains from v to w in $\tilde{D}$, and the capacity of a minimal cut refers to the minimal number of arcs in a vw-disconnecting set of $\tilde{D}$. The result now follows immediately from the integrity theorem (theorem 28c).

The extension of this result to networks in which all the capacities are rational numbers is effected simply by multiplying all these capacities by a suitable integer d to make them integral (e.g. the least common multiple of the denominators of the capacities); we then have the case described in the previous paragraph, and the result follows on dividing by d.

Finally, if some of the capacities are irrational numbers, then the theorem is proved by approximating these capacities as closely as we please by rationals and using the result of the previous paragraph. By carefully choosing these rationals, we can always ensure that the value of any maximal flow and the capacity of any minimal cut are each altered by an amount which we can make as small as we wish. The precise details of this argument will be left as an exercise for the reader. In practical examples, of course, such irrational capacities would rarely occur since the capacities would generally be given in decimal form.//

★*Second proof.* We now give a direct proof of the max-flow min-cut theorem. Note that since the value of any maximal flow cannot exceed the capacity of any minimal cut, it will be sufficient to prove the existence of a cut whose capacity is equal to the value of a given maximal flow.

Let ϕ be a maximal flow. We define two sets V and W of vertices of the network as follows: if G denotes the underlying graph of the digraph D of the network, then a vertex z of the network is contained in V if and only if there exists in G a chain $v = v_0 \to v_1 \to v_2 \to \ldots \to v_{m-1} \to v_m = z$, with the property that each edge $\{v_i, v_{i+1}\}$ corresponds either to an arc (v_i, v_{i+1}) which is unsaturated, or to an arc (v_{i+1}, v_i) which carries a non-zero flow. (Note that v is trivially contained in V). The set W then consists of all those vertices which do not lie in V. For example, in Fig. 29.2, the set V consists of the vertices v, x and y, and the set W consists of the vertices z and w.

We shall now show that W is non-empty, and that in particular it contains the vertex w. If this is not so, then w is in V, and hence there exists in G a chain $v \to v_1 \to v_2 \to \ldots \to v_{m-1} \to w$ of the above type. We now choose a positive number ε satisfying the following two conditions: (*i*) ε must not exceed any of the amounts needed to saturate the arcs of the first type, and (*ii*) ε must not exceed the flow in any of the arcs of the second type. It is now easy to see that if we increase by ε the flow in the arcs of the first type and decrease by ε

the flow in the arcs of the second type, then the effect will be to increase the value of the flow to $\phi + \varepsilon$. But this contradicts our assumption that ϕ is a maximal flow, and it therefore follows that w is contained in W.

To complete the argument, we let E denote the set of all arcs of the form (x, z), where x is in V and z is in W. Clearly E is a cut. Moreover, it is easy to see that every arc (x, z) of E is saturated, since otherwise z would also be an element of V. It follows that the capacity of E must be equal to the value of ϕ, and that E is therefore the required cut.// ★

The max-flow min-cut theorem provides a useful check on the maximality or otherwise of a given flow, as long as the network is fairly simple. In practice, of course, the networks one has to deal with are large and complicated, and it will in general be difficult to find a maximal flow by inspection. We conclude this section with an algorithm for finding a maximal flow in any network with integral capacities; the extension of this algorithm to networks with rational capacities is trivial and will be left to the reader.

Suppose then that we are given a network $N = (D, \psi)$; the finding of a maximal flow in N involves three steps:

Step 1. We first find by inspection a flow ϕ whose value is non-zero (if one exists). For example, if N is the network of Fig. 29.3 then a suitable flow would be the flow shown in Fig. 29.4. It is worth pointing out that the larger we can make the value of our initial flow ϕ, the easier the subsequent steps will be.

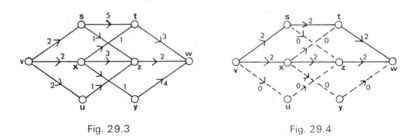

Fig. 29.3   Fig. 29.4

Step 2. We next construct from N a new network N' obtained by reversing the direction of the flow ϕ. More precisely, any arc a for which $\phi(a) = 0$ appears in N' with its original capacity, but any arc a for which $\phi(a) \neq 0$ is replaced by an arc a with capacity

$\psi(a) - \phi(a)$ together with an arc in the direction opposite to a with capacity $\phi(a)$. In our particular example, the network N' takes the form shown in Fig. 29.5; note that v is no longer a source and w no longer a sink.

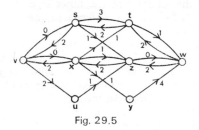

Fig. 29.5

Step 3. If in the network N' we can find a non-zero flow from v to w, then this flow can be added to the original flow ϕ to give a flow ϕ' of larger value in N; we can now repeat step 2 using our new flow ϕ' in place of ϕ in the construction of the network N'. On continuing this procedure, we will eventually end up with a network N' which contains no non-zero flow; the corresponding flow ϕ will then be a maximal flow, as the reader can easily show. In Fig. 29.5, for example,

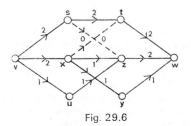

Fig. 29.6

there is a non-zero flow in which the flow in the arcs (v, u), (u, z), (z, x), (x, y) and (y, w) is one and in the remaining arcs is zero. Adding this to the flow of Fig. 29.4 results in the flow shown in Fig. 29.6, which may easily be shown to be maximal by repeating step 2. We have thus obtained the required maximal flow.

In this section we have been able only to scratch the surface of this very diverse and important subject; the reader who wishes to pursue these topics further should consult Ford and Fulkerson.[11]

F

Exercises

(*29a*) Show that the flows of Figs. 29.2 and 29.6 are respectively maximal flows for the networks of Figs. 29.1 and 29.3, and verify the max-flow min-cut theorem in each case.

(*29b*) Show how the analysis of the flows in a network with several sources and sinks can be reduced to the standard case by the introduction of two extra vertices. How would you reduce to the standard case a network problem in which (*i*) some of the arcs are replaced by edges in which the flow can be in either direction, and (*ii*) there are 'capacities' associated with some of the vertices, giving the maximum flow which can pass through those vertices?

(*29c*) Verify the max-flow min-cut theorem for the network of Fig. 29.7 (the omission of an arrow on an arc indicates that the flow is allowed in either direction).

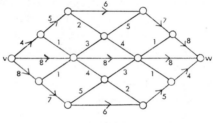

Fig. 29.7

(*29d*) Obtain an algorithm for finding the maximum number of arc-disjoint dichains connecting two given vertices in a digraph.

(**29e*) Suppose that the flow in every arc of a network is bounded below instead of being bounded above (by the capacity); obtain a corresponding 'min-flow max-cut' theorem. What can you say if the flow in every arc is bounded both above and below?

(**29f*) Show how the max-flow min-cut theorem can be used to prove (*i*) Hall's theorem; (*ii*) theorem 27D on the existence of common transversals; (*iii*) theorem 23A on Eulerian digraphs.

(**29g*) Suppose that the numbers in Fig. 29.7 refer to the distances between the corresponding vertices; find the shortest distance from *v* to *w*. (Hint: Consider Fig. 29.7 as a planar graph *G* and form its dual *G**, giving each edge of *G** the same capacity as the corresponding edge in *G*. Now apply the max-flow min-cut theorem to *G**.)

9 Matroid Theory

*All generalizations are
dangerous, even this one.*
ALEXANDRE DUMAS (fils)

In this chapter we shall investigate the rather unexpected similarity
between certain results in graph theory and their analogues in
transversal theory (for example, exercises *9g* and *26e*, or exercises
9j and *26f*). In order to do this it is convenient to introduce the idea
of a matroid, which is essentially a set with an 'independence structure' defined on it. As we shall see, the notion of independence
generalizes not only that of independence in graphs (as defined in
exercise *5j*) but also that of linear independence in vector spaces;
the link with transversal theory is then provided by exercise *26e*.
In §32 we shall show how to define duality in matroids in such a way
as to explain the similarity between the properties of circuits and
cutsets in a graph; it will follow from this that the rather unintuitive
definitions of an abstract-dual and a Whitney-dual of a graph
(§§15–16) arise as natural consequences of matroid duality. In the
final section, we shall show how matroids can be used to give
'easy' proofs of results in transversal theory, and will conclude with
matroid proofs of two deep results in graph theory.

§30. INTRODUCTION TO MATROIDS

In §9 we defined a spanning tree in a connected graph G to be a
connected subgraph of G which contains no circuits and which
includes every vertex of G. It is clear that a spanning tree cannot
contain any other spanning tree as a proper subgraph. It can also
be shown (see exercise *9g*) that if B_1 and B_2 are spanning trees of
G and e is any edge of B_1, then we can find an edge f in B_2 with the
property that $(B_1 - \{e\}) \cup \{f\}$ (i.e. the graph obtained from B_1 on
replacing e by f) is also a spanning tree of G.

Analogous results hold also in the theory of vector spaces and in
transversal theory. If V is a vector space and B_1 and B_2 are bases of

V, then given any element e of B_1, we can find an element f of B_2 with the property that $(B_1 - \{e\}) \cup \{f\}$ is also a basis of V; the corresponding result in transversal theory appears in exercise 26e. Using these three examples as motivation, we can now give our first definition of a matroid.

A **matroid** M is a pair $(E, \mathscr{B})$, where E is a non-empty finite set and $\mathscr{B}$ is a non-empty collection of subsets of E (called **bases**) satisfying the following properties:

($\mathscr{B}$ i) no base properly contains another base;

($\mathscr{B}$ ii) if B_1 and B_2 are bases and if e is any element of B_1, then there is an element f of B_2 with the property that $(B_1 - \{e\}) \cup \{f\}$ is also a base.

By repeatedly using property ($\mathscr{B}$ ii), it is a straightforward exercise to show that any two bases of a matroid M contain the same number of elements; this number is called the **rank** of M.

As we indicated above, a matroid can be associated in a natural way with any graph G by letting E be the set of edges of G and taking as bases the edges of the spanning forests of G; for reasons which will appear later, this matroid is called the **circuit matroid** of G and is denoted by $M(G)$. Similarly, if E is a finite set of vectors in a vector space V, then we can define a matroid on E by taking as bases all linearly independent subsets of E which span the same subspace as E; a matroid obtained in this way is called a **vector matroid**. We shall consider such matroids in further detail later.

A subset of E will be called **independent** if it is contained in some base of the matroid M. It follows that the bases of M are precisely the maximal independent sets, i.e. those independent sets which are contained in no larger independent set, and hence that any matroid is uniquely defined by specifying its independent sets. In the case of a vector matroid, a subset of E is independent if and only if its elements are linearly independent when regarded as vectors in the vector space; similarly, if G is a graph, then the independent sets of $M(G)$ are simply those sets of edges of G which contain no circuit, in other words the edge-sets of the forests contained in G.

Since a matroid can be completely described by listing its independent sets, it seems reasonable to ask whether there is a simple definition of a matroid in terms of its independent sets. One such definition will now be given; the interested reader will find a proof of the equivalence of this definition and the above one in Whitney.[15]

A **matroid** is a pair $(E, \mathscr{I})$, where E is a non-empty finite set, and $\mathscr{I}$ is a non-empty collection of subsets of E (called **independent sets**) satisfying the following properties:

($\mathscr{I}$ i) any subset of an independent set is independent;

($\mathscr{I}$ ii) if I and J are independent sets containing k and $k+1$ elements respectively, then there is an element e contained in J but not in I, such that $I \cup \{e\}$ is independent.

(Note that with this definition, a **base** is defined to be any maximal independent set; property ($\mathscr{I}$ ii) can then be used repeatedly to show that any independent set can be extended to a base.)

If $M = (E, \mathscr{I})$ is a matroid defined in terms of its independent sets, then a subset of E is said to be **dependent** if it is not independent; a minimal dependent set is called a **circuit**. Note that if $M(G)$ is the circuit matroid of a graph G, then the circuits of $M(G)$ are precisely the circuits of G. It is clear that since a subset of E is independent if and only if it contains no circuits, a matroid can be defined in terms of its circuits; one such definition, generalizing to matroids the result of exercise *5f*, is given in exercise *30e*.

Before proceeding to some examples of matroids, it will be convenient to give one further definition of a matroid; this definition, in terms of a **rank function** ρ, is essentially the one given by Whitney in his pioneering paper of 1935.[15]

If $M = (E, \mathscr{I})$ is a matroid defined in terms of its independent sets, and if A is a subset of E, then the size of the largest independent set contained in A is called the **rank** of A and is denoted by $\rho(A)$; note that the previously-defined rank of M is then equal to $\rho(E)$. Since a subset A of E is independent if and only if $\rho(A) = |A|$, it follows that a matroid may be defined in terms of its rank function, as we now show.

THEOREM 30A. *A matroid may be defined as a pair* (E, ρ), *where* E *is a non-empty finite set, and* ρ *is an integer-valued function defined on the set of subsets of* E *and satisfying*:

(ρ i) $0 \leqq \rho(A) \leqq |A|$, *for every subset* A *of* E;

(ρ ii) *if* $A \subseteq B \subseteq E$, *then* $\rho(A) \leqq \rho(B)$;

(ρ iii) *for any* $A, B \subseteq E$, $\rho(A \cup B) + \rho(A \cap B) \leqq \rho(A) + \rho(B)$.

Remark. Note that this is the extension to matroids of the results of exercises *9j* and *26f*.

Proof. We assume first that $M = (E, \mathscr{I})$ is a matroid defined in terms of its independent sets; we wish to prove properties (ρi)—(ρiii).

Clearly (ρ i) and (ρ ii) are trivial. To prove (ρ iii), we let X be a base (i.e. a maximal independent subset) of $A \cap B$; since X is an independent subset of A, X can be extended to a base Y of A, and then (in a similar way) to a base Z of $A \cup B$. Since $X \cup (Z - Y)$ is clearly an independent subset of B, it follows that

$$\rho(B) \geqq \rho(X \cup (Z - Y)) = |X| + |Z| - |Y|$$

$$= \rho(A \cap B) + \rho(A \cup B) - \rho(A),$$

as required.

Conversely, let $M = (E, \rho)$ be a matroid defined in terms of a rank function ρ, and define a subset A of E to be independent if and only if $\rho(A) = |A|$. It is then a straightforward matter to prove property ($\mathscr{I}$ i). To prove ($\mathscr{I}$ ii), let I and J be independent sets containing k and $k+1$ elements respectively, and suppose for every element e which lies in J but not in I that $\rho(I \cup e) = k$. If e and f are two such elements, then

$$\rho(I \cup e \cup f) \leqq \rho(I \cup e) + \rho(I \cup f) - \rho(I) = k;$$

it follows that $\rho(I \cup e \cup f) = k$. We now continue this procedure, adding one new element of J at a time; since at each stage the rank has value k, we conclude that $\rho(I \cup J) = k$, and hence (by ρ ii) that $\rho(J) \leqq k$, which is a contradiction. It follows that there exists an element f which is in J but not in I with the property that $\rho(I \cup f) = k+1.//$

We conclude this section with two simple definitions. A **loop** of a matroid $M = (E, \rho)$ is an element e of E satisfying $\rho(\{e\}) = 0$, and a **pair of parallel elements** of M is a pair $\{e, f\}$ of elements of E which are not loops and which satisfy $\rho(\{e, f\}) = 1$. The reader should verify that if M is the circuit matroid of a graph G, then the loops and parallel elements of M correspond to loops and multiple edges of G.

Exercises

(30a)   Show that any two bases of a matroid on a set E contain the same number of elements; show moreover that if A is any subset of E, then any two maximal independent subsets of A contain the same number of elements.

(*30b*) Use exercise *9j* to show that the rank function of the circuit matroid of a graph G is precisely the cutset rank κ.

(*30c*) Let E be a finite non-empty set, and $\mathscr{S}$ be a family of non-empty subsets of E; show that the partial transversals of $\mathscr{S}$ are the independent sets of a matroid on E. Deduce the result of exercise *26d* (see also §**33**).

(*30d*) Let $M = (E, \rho)$ be a matroid with rank function ρ; show that $M^* = (E, \rho^*)$ is also a matroid, where ρ^* is defined by $\rho^*(A) = |A| + \rho(E - A) - \rho(E)$. Show also that B is a base of M if and only if $E - B$ is a base of M^*. (This exercise will be solved in §**32**).

(**30e*) Show that a matroid M can be defined as a pair $(E, \mathscr{C})$, where E is a non-empty finite set, and $\mathscr{C}$ is a collection of subsets of E (called **circuits**) satisfying the following properties: (*i*) no circuit properly contains another circuit; (*ii*) if C_1 and C_2 are two distinct circuits each containing an element e, then there exists a circuit in $C_1 \cup C_2$ which does not contain e.

(**30f*) Show that the cutsets of a graph G satisfy the conditions of the previous exercise; what is the relationship between the rank function of the corresponding matroid and the rank function of the circuit matroid of G?

(*30g*) If $M = (E, \rho)$ is a matroid, then the **closure** $\phi(A)$ of a subset A of E is the set of all elements e of E with the property that $\rho(A \cup \{e\}) = \rho(A)$; show that (*i*) $A \subseteq \phi(A) = \phi(\phi(A))$; (*ii*) if $A \subseteq B \subseteq E$ then $\phi(A) \subseteq \phi(B)$; (*iii*) if e is contained in $\phi(A \cup \{f\})$ but not in $\phi(A)$, then f is contained in $\phi(A \cup \{e\})$. Defining a set A to be **closed** if $\phi(A) = A$, show that the intersection of two closed sets is a closed set; does an analogous result hold for the union of two closed sets?

(**30h*) Let $M_1 = (E, \mathscr{I}_1)$ and $M_2 = (E, \mathscr{I}_2)$ be two matroids defined on the same set E; show that the set of all unions $I \cup J$ of an independent set I of M_1 and an independent set J of M_2 form the independent sets of a new matroid. (This matroid is called the **union** of M_1 and M_2 and is denoted by $M_1 \cup M_2$; we shall calculate its rank function in §**33**.)

(*30i*) Show how the definition of a fundamental system of circuits in a graph may be extended to matroids; how would you extend the definition of a fundamental system of cutsets?

(**30j*) Show how the definition of a matroid may be extended to infinite sets E, and investigate the properties of such matroids.

§31. EXAMPLES OF MATROIDS

In this section we shall examine several important types of matroid.

TRIVIAL MATROIDS. Given any non-empty finite set E, we can define on it a matroid whose only independent set is the empty set. This matroid is called the **trivial matroid** on E, and is clearly a matroid of rank zero.

DISCRETE MATROIDS. At the other extreme is the **discrete matroid** on E, in which every subset of E is independent; note that the discrete matroid on E has only one base, namely E itself, and that the rank of any subset A is simply the number of elements in A.

UNIFORM MATROIDS. Both of the previous examples are special cases of the **k-uniform matroid** on E, whose bases are those subsets of E which contain exactly k elements. It follows that the independent sets are those subsets of E containing not more than k elements, and that the rank of any subset A is either $|A|$ or k, whichever is smaller. Note that the trivial matroid on E is 0-uniform and that the discrete matroid is $|E|$-uniform.

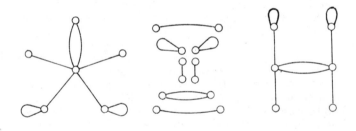

Fig. 31.1

Before developing the examples described in the previous section, it will be convenient to formalize the idea of isomorphism between matroids. Two matroids $M_1 = (E_1, \mathscr{I}_1)$ and $M_2 = (E_2, \mathscr{I}_2)$ are said to be **isomorphic** if there is a one-one correspondence between the sets E_1 and E_2 which preserves independence—in other words, a set of elements of E_1 is independent in M_1 if and only if the corresponding set of elements of E_2 is independent in M_2. As an example, note that the circuit matroids of the three graphs in Fig. 31.1 are all isomorphic—we emphasize the fact that although matroid isomorphism preserves circuits, cutsets and the number of edges in a graph, it does *not* in general preserve connectedness, the number of vertices, or their degrees. Using the above definition of isomorphism, we can now define graphic, transversal and representable matroids.

GRAPHIC MATROIDS. As we saw in the previous section, we can define a matroid $M(G)$ on the set of edges of a graph G by taking the circuits of G as the circuits of the matroid; $M(G)$ is then called the circuit matroid of G and its rank function is simply the cutset rank κ (see exercise *30b*). It is a reasonable question to ask whether a given matroid M is the circuit matroid of some graph—in other words, whether there exists a graph G such that M is isomorphic to $M(G)$; such matroids are called **graphic matroids,** and a characterization of them will be given in the next section. As an example of a graphic matroid, consider the matroid M on the set $\{1, 2, 3\}$ whose independent sets are $\varnothing$, $\{1\}$, $\{2\}$, $\{3\}$, $\{1, 2\}$, and $\{1, 3\}$; clearly M is isomorphic to the circuit matroid of the graph shown in Fig. 31.2.

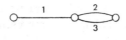

Fig. 31.2

It can be shown, however, that non-graphic matroids exist; the simplest example is the 2-uniform matroid on a set of four elements, as the reader may easily verify.

COGRAPHIC MATROIDS. Given a graph G, the circuit matroid $M(G)$ is not the only matroid which can be defined on the set of edges of G. Because of the similarity between the properties of circuits and of cutsets in a graph, it is reasonable to hope that a matroid can be constructed by taking the cutsets of G as circuits of the matroid. We saw in exercise *30f* that this construction does in fact define a matroid, and we shall refer to it as the **cutset matroid** of G, written $M^*(G)$; note that a set of edges of G is independent if and only if it contains no cutset of G. We shall call a matroid M **cographic** if there exists a graph G such that M is isomorphic to $M^*(G)$; the reason for the name 'cographic' will appear later.

PLANAR MATROIDS. A matroid which is both graphic and cographic is called a **planar matroid;** we shall indicate the connexion between planar matroids and planar graphs in the next section.

REPRESENTABLE MATROIDS. Since the definition of a matroid is partly motivated by the idea of linear independence in vector spaces, it is of interest to investigate those matroids which arise as vector matroids associated with some set of vectors in a vector space over a given field. More precisely, given a matroid M on a set E, we shall say that M is **representable over a field F** if there exist a vector space V over F and a map ϕ from E to V, with the property that a subset A of E is independent in M if and only if ϕ is one-one on A and $\phi(A)$ is linearly independent in V. (Note that this amounts to saying that if we ignore loops and parallel elements then M is isomorphic to a vector matroid defined in some vector space over F.) Of particular importance are those matroids which are representable over the field of integers modulo two—such matroids will be called **binary matroids.** For convenience, we often say simply that M is a **representable matroid** if there exists some field F such that M is representable over F. It turns out that some matroids are representable over every field (the so-called **regular matroids**), some representable over no field and some representable only over some restricted class of fields.

It is not difficult to show that if G is a graph, then its circuit matroid $M(G)$ is a binary matroid. To see this, we associate with each edge of G the corresponding row in the incidence matrix of G (see exercise *2f*), regarded as a vector each of whose components is zero or one. Note that if a set of edges of G form a circuit, then the sum (modulo two) of the corresponding vectors is zero.

An example of a binary matroid which is neither graphic nor cographic will be given in exercise *31i*.

TRANSVERSAL MATROIDS. Our next example provides the link between matroid theory and transversal theory. We recall from exercises *26e*, *26f* and *30c* that if E is a non-empty finite set and $\mathscr{S} = (S_1, \ldots, S_m)$ is a family of non-empty subsets of E, then the partial transversals of $\mathscr{S}$ may be taken as the independent sets of a matroid on E. Any matroid obtained in this way (for a suitable choice of E and $\mathscr{S}$) is called a **transversal matroid** and is denoted by $M(S_1, \ldots, S_m)$. For example, the graphic matroid M described above is a transversal matroid on the set $\{1, 2, 3\}$, since its independent sets are the partial transversals of the family $\mathscr{S} = (S_1, S_2)$, where $S_1 = \{1\}$ and $S_2 = \{2, 3\}$. Note that the rank of a subset A of E is the size of the largest partial transversal contained in A. An example of a matroid which is not transversal will be given in exercise *31d*.

It has recently been proved that every transversal matroid is representable over some field, but is binary if and only if it is graphic. Further results on transversal matroids will be discussed in §33.

RESTRICTIONS AND CONTRACTIONS. In graph theory it is often possible to investigate the properties of a graph by looking at its subgraphs or by considering the graph obtained by contracting some of its edges; we shall find it useful to define the corresponding notions in matroid theory. If M is a matroid defined on a set E and A is a subset of E, then the **restriction** of M to A (denoted by $M \times A$) is the matroid whose circuits are precisely those circuits of M which are contained in A; similarly we define the **contraction** of M to A (denoted by $M.A$) as the matroid whose circuits are obtained by taking the minimal members of the collection $\{C_i \cap A\}$, where the C_i denote circuits of M. (A simpler definition will be given in exercise $32b$). We leave it to the reader to verify that these are in fact matroids, and that they correspond to the deletion and contraction of edges in a graph. A matroid obtained from M by a succession of restrictions and contractions is called a **minor** of M.

BIPARTITE AND EULERIAN MATROIDS. We conclude this section by showing how bipartite and Eulerian matroids may be defined. Since the usual definitions of bipartite and Eulerian graphs as given in §3 and §6 are unsuitable for matroid generalization, we must find alternative characterizations of these graphs. In the case of bipartite graphs, exercise $5g$ comes to our rescue—a **bipartite matroid** is a matroid, every circuit of which contains an even number of elements; for Eulerian graphs we use corollary 6D and define a matroid on a set E to be an **Eulerian matroid** if E can be expressed as the union of disjoint circuits. In the next section we shall see that Eulerian matroids and bipartite matroids are (in a sense to be made precise) dual concepts.

Exercises

($31a$) Show that up to isomorphism there are exactly four matroids on a set of two elements and eight matroids on a set of three elements; how many are there on a set of four elements? (See the appendix.)

($31b$) Show that up to isomorphism the number of matroids on a set of n elements is at most 2^{2^n}, and the number of transversal matroids is at most 2^{n^2}.

(*31c*) Show that every k-uniform matroid is a transversal matroid.

(*31d*) Show that the circuit matroid of K_4 is not a transversal matroid, and find another non-transversal matroid on a set of six elements.

(*31e*) Let M be a matroid of rank r on a set of n elements, and let b and c denote the number of bases and circuits of M; show that

$$b \leqq \binom{n}{r} \quad \text{and} \quad c \leqq \binom{n}{r+1}.$$

(*31f*) Show that the circuit matroids $M(K_5)$ and $M(K_{3,3})$ are both graphic but not cographic; find two matroids which are cographic but not graphic.

(*31g*) Let M be a matroid on a set E, and let $A \subseteq B \subseteq E$; show that (*i*) $(M \times B) \times A = M \times A$; (*ii*) $(M.B).A = M.A$; (*iii*) $(M.B) \times A = (M \times (E-(B-A))).A$; (*iv*) $(M \times B).A = (M.(E-(B-A))) \times A$.

(*31h*) Show that if M satisfies any of the following properties, then so does any minor of M: (*i*) graphic; (*ii*) cographic; (*iii*) binary; (*iv*) regular.

(**31i*) The **Fano matroid** F is the matroid defined on the set $E = \{1, 2, 3, 4, 5, 6, 7\}$ whose bases are all those subsets of E containing three elements except $\{1, 2, 3\}$, $\{1, 4, 5\}$, $\{1, 6, 7\}$, $\{2, 4, 7\}$, $\{2, 5, 6\}$, $\{3, 4, 6\}$ and $\{3, 5, 7\}$; show that F may be drawn as in Fig. 31.3,

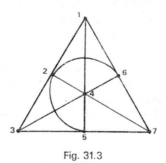

Fig. 31.3

the bases being precisely those sets of three elements which do not lie on a line. Show also that F is (*i*) binary; (*ii*) non-regular; (*iii*) non-transversal; (*iv*) neither graphic nor cographic; (*v*) Eulerian.

(**31j*) If M is a matroid on a set E, then a non-trivial subset A of E is said to be a **separating subset** if $M \times A = M.A$; show that the following conditions are equivalent: (*i*) A is a separating subset; (*ii*) every circuit of M is contained either in A or in $E-A$; (*iii*) $\rho(A) + \rho(E-A) = \rho(E)$. What can you say about a graph whose circuit matroid contains no separating subset?

(*31k) Let D be a digraph without loops, and let E and Y be two disjoint sets of vertices of D. A subset A of E is called **independent** if there exist $|A|$ vertex-disjoint dichains from A to Y. Show that these independent sets form the independent sets of a matroid on E, and show also that every transversal matroid can be obtained in this way. (Such a matroid is called a **gammoid**).

§32. MATROIDS AND GRAPH THEORY

We come now to a study of duality in matroids, our aim being to show how several of the results which appeared earlier in the book seem far more natural when looked at in this light. We shall see, for example, that the rather artificial definitions of an abstract-dual and a Whitney-dual of a planar graph (see §§15, 16) arise as direct consequences of the corresponding definition of a matroid-dual. The point we shall be trying to get across is that not only do various concepts in matroid theory generalize their counterparts in graph theory—they frequently simplify them as well.

We start by recalling from our examination of cographic matroids that we can form a matroid $M^*(G)$ on the set of edges of a graph G by taking as circuits of $M^*(G)$ the cutsets of G; in view of theorem 15c it would seem sensible to choose our definition of the dual of a matroid in such a way as to make this matroid the dual of the circuit matroid $M(G)$ of G.

This may be achieved as follows: if $M = (E, \rho)$ is a matroid defined in terms of its rank function, we define the **dual matroid** of M (denoted by M^*) to be the matroid on E whose rank function ρ^* is given by

$$\rho^*(A) = |A| + \rho(E-A) - \rho(E), \qquad (A \subseteq E).$$

We must first verify that ρ^* actually is the rank function of a matroid on E.

THEOREM 32A. $M^* = (E, \rho^*)$ *is a matroid on* E.

Proof. We must verify the properties $(\rho\ i)$, $(\rho\ ii)$, and $(\rho\ iii)$ of §30, for the function ρ^*.

To prove $(\rho\ i)$, we note first that $\rho(E-A) \leq \rho(E)$, and hence that $\rho^*(A) \leq |A|$. Also, (by $(\rho\ iii)$ applied to the function ρ) we have $\rho(E) + \rho(\varnothing) \leq \rho(A) + \rho(E-A)$, and hence that $\rho(E) - \rho(E-A) \leq \rho(A) \leq |A|$. It follows immediately that $\rho^*(A) \geq 0$. The proof of $(\rho\ ii)$ is equally straightforward, and will be left as an exercise.

To prove (ρ iii), we have, for any $A, B \subseteq E$,

$$\rho^*(A \cup B) + \rho^*(A \cap B) = |A \cup B| + |A \cap B| + \rho(E - (A \cup B)) \\ + \rho(E - (A \cap B)) - 2\rho(E)$$

$$= |A| + |B| + \rho((E - A) \cap (E - B)) \\ + \rho((E - A) \cup (E - B)) - 2\rho(E)$$

$$\leqq |A| + |B| + \rho(E - A) + \rho(E - B) - 2\rho(E) \\ \text{(by (ρ iii), applied to ρ).}$$

$$= \rho^*(A) + \rho^*(B), \text{ as required.} //$$

Although the above definition seems highly contrived, it turns out that the bases of M^* can be described very simply in terms of the bases of M, as we now show:

THEOREM 32B. *The bases of M^* are precisely the complements of the bases of M.*

Remark. This result can, in fact, be used to define M^*.

Proof. We shall show that if B^* is a base of M^*, then $E - B^*$ is a base of M; the converse result is obtained by simply reversing the argument.

Since B^* is independent in M^*, $|B| = \rho^*(B^*)$, and hence $\rho(E - B^*)$ $= \rho(E)$. It thus remains only to prove that $E - B^*$ is independent in M. But this follows immediately from the fact that $\rho^*(B) = \rho^*(E)$, on using the above expression for ρ^*.//

As an immediate consequence of the above definition, we observe that (in contrast to the duality of planar graphs) every matroid has a dual and this dual is unique. It also follows immediately from theorem 32B that the double-dual M^{**} is equal to M; in fact, as we shall see, this completely trivial result is the natural generalization to matroids of the (non-trivial) results of theorems 15B and 15E and corollary 16B.

We shall now show that the cutset matroid $M^*(G)$ of a graph G is the dual of the circuit matroid $M(G)$:

THEOREM 32C. *If G is a graph, then $M^*(G) = (M(G))^*$.*

Proof. Since the circuits of $M^*(G)$ are the cutsets of G, we must check that C^* is a circuit of $(M(G))^*$ if and only if C^* is a cutset of G.

Suppose first that C^* is a cutset of G. If C^* is independent in $(M(G))^*$, then C^* can be extended to a base B^* of $(M(G))^*$. It follows that $C^* \cap (E - B^*)$ is empty, contradicting the result of theorem 9c since $E - B^*$ is a spanning forest of G. It follows that C^* is a dependent set in $(M(G))^*$, and thus contains a circuit of $(M(G))^*$.

If, on the other hand, D^* is a circuit of $(M(G))^*$, then D^* is not contained in any base of $(M(G))^*$. It follows that D^* intersects every base of $M(G)$, i.e. every spanning forest of G. Hence, by the result of exercise *9h*, D^* contains a cutset. The result now follows.//

Before proceeding further, it will be convenient to introduce some more terminology. We shall say that a set of elements of a matroid M form a **cocircuit** of M if they form a circuit of M^*; note that in view of theorem 32c the cocircuits of the circuit matroid of a graph G are precisely the cutsets of G. We can similarly define a **cobase** of M to be a base of M^*, with corresponding definitions for **corank, co-independent set,** etc. We shall also say that a matroid M is **cographic** if and only if its dual M^* is graphic, and in view of theorem 32c this definition agrees with the one given in the previous section. The reason for introducing this 'co-notation' is that we may now restrict ourselves to a single matroid M without having to bring in M^*. To illustrate this, we shall prove the analogue for matroids of theorem 9c.

THEOREM 32D. *Every cocircuit of a matroid intersects every base.*

Proof. Let C^* be a cocircuit of a matroid M, and suppose that there exists a base B of M with the property that $C^* \cap B$ is empty. Then C^* is contained in $E - B$, and so C^* is a circuit of M^* which is contained in a base of M^*; this contradiction establishes the result.//

COROLLARY 32E. *Every circuit of a matroid intersects every cobase.*
Proof. Apply the result of theorem 32D to the matroid M^*.//

Note that by taking a matroid point of view, the two results in theorem 9c turn out to be dual forms of a single result; thus, instead of proving two results in graph theory (as we had to in §9), it is sufficient to prove a single result in matroid theory and then use duality. Not only does this represent a considerable saving of time and effort, it also gives us greater insight into several of the problems we have encountered earlier in the book. One example of this is the often-mentioned similarity between the properties of circuits and cutsets; another is a deeper understanding of duality in planar graphs.

As a further example of the simplification introduced by matroid theory, let us look again at exercise *5f*. A straightforward proof of this result would involve two separate operations—a proof for circuits and a different proof for cutsets; if, however, we prove the matroid analogue of the result for circuits (as stated in exercise *30e*), then we can simply apply it to the matroid $M^*(G)$, and immediately obtain the corresponding result for cutsets; conversely, we can use duality to deduce the result for circuits from the result for cutsets.

Let us now turn our attention to planar graphs, and in particular to the problem of showing how the definitions of a geometric-dual, an abstract-dual and a Whitney-dual of a graph all arises as consequences of duality in matroids. We shall start with the abstract-dual:

THEOREM 32F. *If G^* is an abstract-dual of a graph G, then $M(G^*)$ is isomorphic to $(M(G))^*$.*

Proof. Since G^* is an abstract-dual of G, there is a one-one correspondence between the edges of G and those of G^* with the property that circuits in G correspond to cutsets in G^* and conversely. It follows immediately from this that the circuits of $M(G)$ correspond to the cocircuits of $M(G^*)$, and hence, by theorem 32C, that $M(G^*)$ is isomorphic to $M^*(G)$, as required.//

COROLLARY 32G. *If G^* is a geometric-dual of a connected planar graph G, then $M(G^*)$ is isomorphic to $(M(G))^*$.*

Proof. This result follows immediately from theorems 32F and 15C.//

Note that (as remarked before) a planar graph can have several different duals, whereas a matroid can have only one; the reason for this is the easily-checked fact that if G^* and G^x are two (possibly non-isomorphic) duals of G, then the circuit matroids of G^* and G^x are isomorphic matroids. We come now to the Whitney-dual of a graph:

THEOREM 32H. *If G^* is a Whitney-dual of a graph G, then $M(G^*)$ is isomorphic to $(M(G))^*$.*

Proof. Since, by the proof of theorem 16C, G^* is a Whitney-dual of G if and only if G^* is an abstract-dual of G, the result follows immediately from theorem 32F. Alternatively, it may readily be shown that the defining equation for the Whitney-dual can be deduced from the formula for ρ^* given in corollary 32B. The details of this argument are left to the reader.//

We conclude this section by giving an answer to the question, 'under what conditions is a given matroid M graphic?' It is not difficult to find necessary conditions. For example, it follows from our discussion of representable matroids (§31) that such a matroid must be binary. Furthermore, by exercises $31f$ and $31i$, it is clear that M cannot contain as a minor any of the matroids $M^*(K_5)$, $M^*(K_{3,3})$, F or F^* (where F denotes the Fano matroid). It was shown in the following deep theorem by Tutte that these necessary conditions are in fact sufficient; the proof of this result is too difficult to be given here.

THEOREM 32I (Tutte 1959). *A matroid M is graphic if and only if it is binary and contains no minor isomorphic to $M^*(K_5)$, $M^*(K_{3,3})$, F or F^*.//*

On applying theorem 32I to M^*, and using the fact (see exercise $32c$) that the dual of a binary matroid is binary, we immediately obtain necessary and sufficient conditions for a matroid to be cographic.

COROLLARY 32J. *A matroid M is cographic if and only if it is binary and contains no minor isomorphic to $M(K_5)$, $M(K_{3,3})$, F or F^*.//*

Tutte also proved that *a binary matroid is regular if and only if it contains no minor isomorphic to F or F^**; by combining this result with the results of theorem 32I and corollary 32J, we immediately deduce the following matroid analogue of theorem 12C.

THEOREM 32K. *A matroid is planar if and only if it is regular and contains no minor isomorphic to $M(K_5)$, $M(K_{3,3})$ or their duals.//*

Exercises

(*32a*) Show by an example that the dual of a transversal matroid is not necessarily transversal.

(*32b*) Show that if M is a matroid on a set E and A is a subset of E, then the contraction $M.A$ is the matroid whose cocircuits are precisely those cocircuits of M which are contained in A. Show also that $(M.A)^* = M^* \times A$ and $(M \times A)^* = M^*.A$, and deduce that if A is a separating subset of M, then A is a separating subset of M^*, and conversely.

(*32c*) Show that for any circuit C and any cocircuit C^* of a matroid M, $|C \cap C^*| \neq 1$. (This is the generalization to matroids of the result of exercise $5i$.) Show moreover that M is binary if and only if $|C \cap C^*|$ is even, and deduce that the dual of a binary matroid is binary.

(*32d) Let M be a binary matroid on a set E. Use the result of the previous exercise to show that if M is an Eulerian matroid then M^* is bipartite; show also (by using induction on $|E|$) that the converse result is true. By considering the 5-uniform matroid on a set of eleven elements, show that the condition that M is binary cannot be omitted. (This exercise provides the generalization to matroids of exercise *15g*.)

(*32e) Show how one can imitate exercise (*2j*) to define the **vector space V associated with a matroid M** on a set E. Prove also that if M is binary, then the sum of any two subsets of E may be written as the union of disjoint circuits of M, and deduce that the set of all unions of disjoint circuits of M form a subspace of V (called the **circuit subspace** of M). Use duality to obtain a corresponding result for cocircuits, and compare these results with those of exercises *6h* and *6i*.

(*32f) Show that for binary matroids, the matroid definition of duality generalizes the definition of an algebraic-dual of a graph (as given in exercise *15j*).

§33. MATROIDS AND TRANSVERSAL THEORY

We showed in the previous section that there is a close connexion between results in matroid theory and in graph theory; the connexion between matroid theory and transversal theory will now be described. Our first aim is to show how the proofs of several of the earlier results on transversal theory may be considerably simplified by taking a matroid-theoretic point of view.

The reader will recall that if E is a non-empty finite set and $\mathscr{S} = (S_1, \ldots, S_m)$ is a family of non-empty subsets of E, then the partial transversals of $\mathscr{S}$ may be taken as the independent sets of a matroid on E, denoted by $M(S_1, \ldots, S_m)$; in this matroid, the rank of a subset A of E is simply the size of the largest partial transversal of $\mathscr{S}$ contained in A.

Our first example of the use of matroids in transversal theory is a proof of the result of exercise *26d* that a family $\mathscr{S}$ of subsets of E has a transversal containing a given subset A if and only if (*i*) $\mathscr{S}$ has a transversal, and (*ii*) A is a partial transversal of $\mathscr{S}$. It is clear that both of these conditions are necessary; to prove that they are sufficient, it is enough to observe that since A is a partial transversal of $\mathscr{S}$, A must be an independent set in the transversal matroid M determined by $\mathscr{S}$ and so can be extended to a base of M. Since $\mathscr{S}$ has a transversal, every base of M must be a transversal of $\mathscr{S}$, and the result follows immediately. The reader who has worked through exercise *26d* will realize how much simpler this argument is.

Before showing how matroid theory can be used to simplify the proof of theorem 27D on the existence of a common transversal of two families of subsets of a set E, we shall prove a natural extension to matroids of Hall's theorem. We recall that if $\mathscr{S}$ is a family of subsets of E, then Hall's theorem gave a necessary and sufficient condition for $\mathscr{S}$ to have a transversal; if we also have a matroid structure defined on E, then it is reasonable to ask whether there is a corresponding condition for the existence of an **independent transversal**, i.e. a transversal of $\mathscr{S}$ which is also an independent set in the matroid. The following theorem, known as **Rado's theorem**, answers this question.

THEOREM 33A (Rado 1942). *Let M be a matroid on a set E, and let $\mathscr{S} = (S_1, \ldots, S_m)$ be a family of non-empty subsets of E; then $\mathscr{S}$ has an independent transversal if and only if the union of any k of the subsets S_i contains an independent set of size at least k (for $1 \leq k \leq m$).*

Remark. If M is the discrete matroid on E, then this theorem reduces to Hall's theorem as stated in theorem 26A.

Proof. We shall imitate the proof of theorem 26A. As before, the necessity of the condition is clear, and it is thus sufficient to prove that if one of the subsets (S_1, say) contains more than one element, then we can remove an element from S_1 without altering the condition. By repeating this procedure, we eventually reduce the problem to the case in which each subset contains only one element, the proof then being trivial.

To show the validity of the reduction procedure, we suppose that S_1 contains elements x and y, the removal of either of which invalidates the condition. Then there are subsets A and B of $\{2, 3, \ldots, n\}$ with the property that

$$\rho(\bigcup_{j \varepsilon A} S_j \cup (S_1 - \{x\})) \leq |A| \quad \text{and} \quad \rho(\bigcup_{j \varepsilon B} S_j \cup (S_1 - \{y\})) \leq |B|.$$

But these two inequalities lead to a contradiction, since

$$|A| + |B| + 1 = |A \cup B| + |A \cap B| + 1$$

$$\leq \rho(\bigcup_{j \varepsilon A \cup B} S_j \cup S_1) + \rho(\bigcup_{j \varepsilon A \cap B} S_j) \quad \text{(by the condition)}$$

$$\leq \rho(\bigcup_{j \varepsilon A} S_j \cup (S_1 - \{x\})) + \rho(\bigcup_{j \varepsilon B} S_j \cup (S_1 - \{y\}) \quad \text{(since } |S_1| \geq 2)$$

$$\leq |A| + |B|, \quad \text{(by hypothesis).} //$$

By imitating the proof of corollary 26B, we immediately obtain the following result:

COROLLARY 33B. *With the above notation, $\mathscr{S}$ has an independent partial transversal of size t if and only if the union of any k of the subsets S_i contains an independent set of size at least $k+t-m$.*

We can now give a matroid-theoretic proof of theorem 27D on the existence of a common transversal of two families of subsets of a given set.

THEOREM 27D. *Let E be a non-empty finite set, and let $\mathscr{S} = (S_1, \ldots, S_m)$ and $\mathscr{T} = (T_1, \ldots, T_m)$ be two families of non-empty subsets of E; then $\mathscr{S}$ and $\mathscr{T}$ have a common transversal if and only if, for all subsets A and B of $\{1, 2, \ldots, m\}$,*

$$|(\bigcup_{i\varepsilon A} S_i) \cap (\bigcup_{j\varepsilon B} T_j)| \geqq |A|+|B|-m.$$

Proof. Let $M = (E, \rho)$ be the matroid whose independent sets are precisely the partial transversals of the family $\mathscr{S}$; then $\mathscr{S}$ and $\mathscr{T}$ have a common transversal if and only if $\mathscr{T}$ has an independent transversal. But by theorem 33A, this is so if and only if the union of any k of the sets T_i contains an independent set of size at least k (for $1 \leqq k \leqq m$)—in other words, if and only if the union of any k of the sets T_i contains a partial transversal of $\mathscr{S}$ of size k. The result now follows from corollary 26C.//

Our next application is to the union of matroids. The reader will recall from exercise *30h* that if M_1 and M_2 are two matroids on a set E, then we can define a new matroid $M_1 \cup M_2$ by taking as independent sets all possible unions of an independent set of M_1 and an independent set of M_2. We shall now find the rank of this matroid.

THEOREM 33C. *If ρ_1 and ρ_2 denote the rank functions of M_1 and M_2, then the rank $\rho(E)$ of the union $M_1 \cup M_2$ is given by*

$$\rho(E) = \min_{A \subseteq E} \{\rho_1(A)+\rho_2(A)+|E-A|\}.$$

★*Proof.* We remark first that if A is any subset of E and B is any base of $M_1 \cup M_2$, then clearly

$$\rho(E) = |B| = |B \cap A|+|B \cap (E-A)| \leqq \rho_1(A)+\rho_2(A)+|E-A| .$$

In order to prove the reverse inequality, we let $E = \{e_1, \ldots, e_n\}$ and let F be any set $\{f_1, \ldots, f_n\}$ whose intersection with E is empty; then we can define on the set F a matroid $\tilde{M}_2$ isomorphic to M_2, this being done in the obvious way. It now follows immediately that $M_1 \cup \tilde{M}_2$ is a matroid on $E \cup F$ whose rank function $\tilde{\rho}$ is given by $\tilde{\rho}(A) = \rho_1(A) + \rho_2(A)$ for any subset A of $E \cup F$.

Consider the family $\mathscr{S} = (S_1, \ldots, S_n)$ of subsets of $E \cup F$, where $S_i = \{e_i, f_i\}$; it is clear that the rank in $M_1 \cup M_2$ of any subset B of E is at least t if and only if the subsets $(S_i : e_i \, \varepsilon B)$ have a partial transversal of size t which is independent in $M_1 \cup \tilde{M}_2$. But, by corollary 33B, this is so if and only if the union of any k of these subsets has rank at least $k + t - |B|$ in $M_1 \cup \tilde{M}_2$. It follows that if U denotes such a union, and A denotes the set of corresponding elements of B, then $\rho_1(U) = \rho_1(A)$ and $\rho_2(U) = \rho_2(A)$. Hence B has rank at least t if and only if $\rho_1(A) + \rho_2(A) = \tilde{\rho}(U) \geqq |A| + t - |B|$.

Since the rank of E is less than $\rho(E) + 1$, it follows, on putting $B = E$ and $t = \rho(E) + 1$, that

$$\rho(E) + 1 > \rho_1(A) + \rho_2(A) + |E - A|.$$

The result now follows.//★

This result can easily be extended by induction to the union of k matroids, and the following corollary gives the rank function of such a union.

COROLLARY 33D. *If $M_1, \ldots, M_k$ are matroids on a set E with rank functions $\rho_1, \ldots, \rho_k$ respectively, then the rank function ρ of $M_1 \cup \ldots \cup M_k$ is given by*

$$\rho(X) = \min_{A \subseteq X} \{\rho_1(A) + \ldots + \rho_k(A) + |X - A|\}.$$

Proof. As just mentioned, the extension of theorem 33C from two matroids to k matroids is effected by a simple induction proof. The rank of any subset X of E is then obtained by restricting this union to X, using the easily-proved fact that $(M_1 \cup \ldots \cup M_k) \times X = (M_1 \times X) \cup \ldots \cup (M_k \times X).//$

We conclude this chapter by showing how the result just proved may be used to obtain two deep results in graph theory. To this end, we shall derive some simple consequences of corollary 33D.

COROLLARY 33E. *Let $M = (E, \rho)$ be a matroid; then M contains k disjoint bases if and only if, for any subset A of E,*

$$k \, \rho(A) + |E - A| \geqq k \, \rho(E).$$

Proof. M contains k disjoint bases if and only if the union of k copies of the matroid M has rank at least $k \, \rho(E)$; the result now follows immediately from corollary 33D.//

COROLLARY 33F. *Let $M = (E, \rho)$ be a matroid; then E can be expressed as the union of not more than k independent sets if and only if, for any subset A of E, $k \, \rho(A) \geqq |A|$.*

Proof. In this case, the union of k copies of the matroid M has rank $|E|$. It thus follows immediately from corollary 33D that $k \, \rho(A) + |E - A| \geqq |E|$, as required.//

If we now apply these two last corollaries to the circuit matroid $M(G)$ of a connected graph G, we immediately obtain necessary and sufficient conditions for G to contain k edge-disjoint spanning trees, and for G to split up into k trees. It turns out, in fact, that these results are not at all easy to obtain by more direct methods, and we have thus once again demonstrated the power of the theory of matroids in solving problems in graph theory.

THEOREM 33G. *A connected graph G contains k edge-disjoint spanning trees if and only if, for any subgraph H of G,*

$$k(\kappa(G) - \kappa(H)) \leqq m(G) - m(H),$$

where $m(H)$ and $m(G)$ denote the number of edges of H and G respectively.//

THEOREM 33H. *A connected graph G may be split up into at most k trees if and only if, for any subgraph H of G, $k \, \kappa(H) \geqq m(H)$.//*

Exercises

(*33a*) Show how the Halmos–Vaughan proof of Hall's theorem (see page 117) may be modified to give a proof of theorem 33A.

(*33b*) Show that if $M = M(S_1, \ldots, S_m)$ is a transversal matroid of rank r, then there exist r of the sets S_i, $(S_{i_1}, \ldots, S_{i_r}$, say) such that $M = M(S_{i_1}, \ldots, S_{i_r})$.

(*33c*) Show that a matroid M is a transversal matroid if and only if M can be expressed as the union of matroids of rank one.

(33d) Let $M_1 = (E, \rho_1)$ and $M_2 = (E, \rho_2)$ be matroids on a set E. Show that M_1 and M_2 have a common independent set of size k if and only if, for every subset A of E,

$$\rho_1(A) + \rho_2(E - A) \geqq k.$$

(33e) Dualize the results of theorems 33G and 33H to obtain two further results in graph theory.

(*33f) Use corollary 33F to obtain a condition under which a finite set of vectors in a vector space can be split up into k disjoint linearly independent subsets. Obtain a corresponding result from corollary 33E.

(*33g) Let $\mathscr{S}$ be a family of non-empty subsets of a set E; obtain a condition for $\mathscr{S}$ to contain t disjoint partial transversals of sizes $r_1, \ldots, r_t$.

Postscript

Although we have now almost reached the end of the book, we have by no means reached the end of the subject. It is our hope that many of our readers will wish to continue their graph-theoretic studies, and for this reason, we thought it might be helpful if we suggested possible directions for further reading.

The reader who is interested primarily in 'pure' graph theory should consult the book by Harary[1], which is a mine of information as well as an excellent reference book. Also worth reading are Moon[7] on trees, and Ore[5] on planarity and colouring problems. For a discussion on the various applications of graph theory, the reader might try Busacker & Saaty[2], Berge[3], or Ford & Fulkerson[11]. This last book is a 'must' for anyone who wishes to pursue seriously the subject of network flows. For a general all-round introduction to combinatorial theory, including not only graph theory and network flows but also an attractive account of enumerative techniques, linear programming, and block designs, there is an excellent book by Liu[6]. A detailed treatment of transversal theory, including a lot of material on transversal matroids, will be found in a recent book by Mirsky[10].

Sooner or later (preferably, sooner) the reader will need to refer to mathematical journals rather than to books. There are a large number of journals which frequently include papers in graph theory and related fields, and there are even two, the *Journal of Combinatorial Theory* (Academic Press), and *Discrete Mathematics* (North-Holland), which are aimed at specialists in these fields. Finally we should perhaps mention that at the end of Harary[9] there is a list of over

1700 papers and books on graph theory published before 1968, indexed under their subject matter as well as the author's name; the reader would be well-advised to spend some time browsing through this list.

Here is my journey's end.
WILLIAM SHAKESPEARE (*Macbeth*)

Appendix

This table lists the number of graphs and digraphs of various types with a given number n of vertices. Numbers greater than one million are given to one significant figure.

Types of Graph n	1	2	3	4	5	6	7	8
Simple Graphs	1	2	4	11	34	156	1044	12 346
Connected Simple Graphs	1	1	2	6	21	112	853	11 120
Eulerian Simple Graphs	1	0	1	1	4	8	37	184
Hamiltonian Simple Graphs	1	0	1	3	8	48	?	?
Trees	1	1	1	2	3	6	11	23
Labelled Trees	1	1	3	16	125	1296	16 807	262 144
Connected Simple Planar Graphs	1	1	2	6	20	105	?	?
Simple Digraphs	1	3	16	218	9608	$\sim 2 \times 10^6$	$\sim 9 \times 10^8$	$\sim 2 \times 10^{12}$
Connected Simple Digraphs	1	2	13	199	9364	$\sim 2 \times 10^6$	$\sim 9 \times 10^8$	$\sim 2 \times 10^{12}$
Strongly-connected Simple Digraphs	1	1	5	90	?	?	?	?
Tournaments	1	1	2	4	12	56	456	6880
Matroids (on n elements)	2	4	8	17	38	?	?	?

Bibliography

*Of making many books there is no
end; and much study is a
weariness of the flesh.*

ECCLESIASTES

1   F. HARARY, *Graph theory*, Addison–Wesley, Reading, Mass., 1969.
2   R. G. BUSACKER and T. L. SAATY, *Finite graphs and networks*, McGraw–Hill, New York, 1965.
3   C. BERGE, *The theory of graphs*, Methuen, London, 1962.
4   O. ORE, *Theory of graphs*, Amer. Math. Soc., Providence, Rhode Island, 1962.
5   O. ORE, *The four-color problem*, Academic Press, New York, 1967.
6   C. L. LIU, *Introduction to combinatorial mathematics*, McGraw–Hill, New York, 1968.
7   J. W. MOON, *Counting labelled trees*, Canadian Math. Congress, Montreal, 1970.
8   B. HARRIS (ed.), *Graph theory and its applications*, Academic Press, New York, 1970.
9   F. HARARY (ed.), *Proof techniques in graph theory*, Academic Press, New York, 1969.
10 L. MIRSKY, *Transversal theory*, Academic Press, New York, 1971.
11 L. R. FORD and D. R. FULKERSON, *Flows in networks*, Princeton Univ. Press, Princeton, New Jersey, 1962.
12 L. CARROLL, *Alice's adventures in Wonderland*, first pub. 1865, an edition also published by Collins, London, 1966.
13 T. M. APOSTOL, *Mathematical analysis*, Addison–Wesley, Reading, Mass., 1957.
14 W. FELLER, *An introduction to probability theory and its applications*, 3rd edn, Wiley, New York, 1968.
15 H. WHITNEY, 'On the abstract properties of linear dependence,' *Amer. J. Math.*, 1935, Vol. 57, 509–533.

Index of Definitions

I've got a little list.
W. S. GILBERT

165